ARTICULATION AND INTELLIGIBILITY

© Springer Nature Switzerland AG 2022

Reprint of original edition © Morgan & Claypool 2005

Articulation and Intelligibility

Jont B. Allen

ISBN: 978-3-031-01426-0   Allen, Articulation and Intelligibility (paperback)
ISBN: 978-3-031-02554-9   Allen, Articulation and Intelligibility (e-book)
Library of Congress Cataloging-in-Publication Data

First Edition
10 9 8 7 6 5 4 3 2 1

ARTICULATION AND INTELLIGIBILITY

Jont B. Allen
Beckman Institute for Advanced Science and Technology
University of Illinois at Champaign-Urbana

ABSTRACT

Immediately following the Second World War, between 1947 and 1955, several classic papers quantified the fundamentals of human speech information processing and recognition. In 1947 French and Steinberg published their classic study on the articulation index. In 1948 Claude Shannon published his famous work on the theory of information. In 1950 Fletcher and Galt published their theory of the articulation index, a theory that Fletcher had worked on for 30 years, which integrated his classic works on loudness and speech perception with models of speech intelligibility. In 1951 George Miller then wrote the first book *Language and Communication*, analyzing human speech communication with Claude Shannon's just published theory of information. Finally in 1955 George Miller published the first extensive analysis of phone decoding, in the form of confusion matrices, as a function of the speech-to-noise ratio. This work extended the Bell Labs' speech articulation studies with ideas from Shannon's Information theory. Both Miller and Fletcher showed that speech, as a code, is incredibly robust to mangling distortions of filtering and noise.

Regrettably much of this early work was forgotten. While the key science of information theory blossomed, other than the work of George Miller, it was rarely applied to aural speech research. The robustness of speech, which is the most amazing thing about the speech code, has rarely been studied.

It is my belief (i.e., assumption) that we can analyze speech intelligibility with *the scientific method*. The quantitative analysis of speech intelligibility requires both *science* and *art*. The scientific component requires an error analysis of spoken communication, which depends critically on the use of statistics, information theory, and psychophysical methods. The artistic component depends on knowing how to restrict the problem in such a way that progress may be made. It is critical to tease out the relevant from the irrelevant and dig for the key issues. This will focus us on the decoding of nonsense phonemes with no visual component, which have been mangled by filtering and noise.

This monograph is a summary and theory of human speech recognition. It builds on and integrates the work of Fletcher, Miller, and Shannon. The long-term goal is to develop a quantitative theory for predicting the recognition of speech sounds. In Chapter 2 the theory is developed for *maximum entropy* (MaxEnt) speech sounds, also called *nonsense speech*. In Chapter 3, context is factored in. The book is largely reflective, and quantitative, with a secondary goal of providing an historical context, along with the many deep insights found in these early works.

KEYWORDS

Speech recognition, phone recognition, robust speech recognition, context models, confusion matrix, features, events, articulation index

Contents

Preface

I would like to say a few words on how this monograph came into existence, as well as give some background as to where I believe this research is headed.

It was a wonderful opportunity to be in the Acoustics Research Department of Bell Labs, and to work with such experienced and marvelous people, having an amazing cross section of disciplines. Thirty two years of these associations has had an impact on me. However, it would be wrong to characterize it as an easy ride. When I first came to Bell I was interested in mathematical modeling, and by one way or another, I came to the problem of modeling the cochlea (the inner ear). I chose cochlear modeling because I felt it was a sufficiently complex problem that it would take a long time to solve. This sense was correct. While we know much more about the cochlea today, we do not yet have a solution to this complex organ's secrets.

In 1983, AT&T was divested by a Justice Department law suit. AT&T had been a regulated monopoly, and there were forces in the government that believed this was a bad thing. Perhaps such regulation was a bit too socialistic for some people. Regardless of what they thought, AT&T was not resistant to being broken up, resulting in the creation of the independent "Baby Bells" and the Mother company AT&T, which controlled a majority of the long distance traffic.

Within months of this breakup, forces within AT&T lobbied for outside ventures. I soon found myself part of one of these ventures, making a new "multi-band wide dynamic range compression" hearing aid. That's another story. The short version is that by 1990 the new algorithm and hardware was a clear winner. During this development phase, I learned many important things, and soon became friends with interesting and important people within the hearing aid industry. AT&T was especially close to CUNY and Profs. Harry Levitt and Arthur Boothroyd, and their collaborators. I had been teaching at CUNY, and the AT&T hearing aid venture was soon using CUNY Ph.D. students, who helped with the development and testing of the new AT&T hearing aid. It was an exciting time for all, and everyone had a contribution to make. One of these Ph.D. students was Patricia Jeng (now my wife). Her job was to develop a fitting system we called "Loudness growth in octave bands," or LGOB, conceived by Joe Hall and myself.

Soon after we were married, while working on her Ph.D., Pat read many of the papers of Harvey Fletcher. Now I had heard of this famous guy, Fletcher, but had never actually read even one of his papers. Pat suggested that I read Fletcher and Munson, 1993, "Loudness, Its Definition, Measurement, and Calculation." I was sure I did not want to read some "silly old" paper on loudness, since I knew (boy was I wrong!) that there was no science in such a topic, loudness being a psychological variable and all, and therefore impossible to quantify. Pat persisted, and I finally gave in, just to pacify her, as best I remember.

That paper changed my life. I found the paper to be fascinating, and the more I learned about Fletcher, the more I realized that I was reading the work of a true genius. I was soon on the war-path to read every one of his papers, finding each to be just as good as the last. Soon after, the word got around that I was becoming a bit of a Fletcher expert, and the ASA Speech Committee asked me to edit a reprinting of Fletcher's 1953 book (Fletcher had two Speech and Hearing books, plus a book on Sunday school teaching). The 1953 book was soon reprinted by the Acoustical Society, and it quickly became an instant "best seller," just like it did in 1929 and again 1953.

One of the major problems that Fletcher and his colleagues worked on during his 30-year career at AT&T, as head of the Acoustics Research Dept (the same department that I was in), was his theory called the *articulation index*, which was an interesting and important measure of the intelligibility of nonsense speech sounds. Now I knew nothing of this, and was not even interested in speech perception, but I had made the commitment to myself to read all of Fletcher's papers and books, and so that is what I did. The more I learned of this AI theory, the more was obvious that (1) nobody really understood it, (2) it was important, and (3) Fletcher had done something really important that was, as best I could tell, largely undervalued. I set out to rectify this situation. I halted my research on cochlear modeling, and began trying to understand, in detail, Fletcher's theory of speech recognition. I was on to something new that was exciting: How do humans communicate?

It took me a long time to understand that there was no theory of speech recognition. When I raised the question "What is the theory of human speech recognition?" I got nasty glares. I had learned early in my career that when you on to something good, people get mad at you when you mention it. And people were getting mad. The madder they got, the more interested I was in this strange topic.

Soon I learned of a second genius, George Miller. While I have not been successful (yet) in reading all of Miller's work's, I have certainly read many, as reflected in this monograph. The work of Fletcher and Miller nicely complement each other. It soon became clear that merging the results of Harvey Fletcher's AI, Claude Shannon's information theory, and George Miller's confusion matrix, was a worthy exercise. Yet the more I talked about it, the madder everyone became. Truth be known, I really tried to push the limits (and I frequently went over them). But by doing this I was able to integrate many diverse opinions, and finally converge, I hope, on a reasonable initial theory of human speech recognition, based on this integrated input from a great number on opinions and discussions. Those of you know who you are, and I thank you.

I would not claim this is anything close to a complete theory. It is not. It is, however, a quantitative theory, and it seems to be the best we have at this time. What is it missing? The problem remains unsolved. The basic units of aural speech remain undiscovered. We have yet to "capture" an event. However I suspect that we are closing in. The scientific method is leading us to this goal, of identifying a perceptual event, and proving, beyond questionable doubt, that events represent what humans extract when they decode speech. No machine today is even close in performing with the accuracy of the human listener. Computers can do many amazing things, but to date they are very poor at speech recognition. The problem is that we do not understand exactly what it is that humans do. Once this problem is solved, I predict that machines will become as good as the very best humans at decoding nonsense sounds in noise. However this still will not solve the problem of matching human performance, since humans are amazing in their ability for extracting subtle cues from context. Machines are notoriously bad at doing this, as well. That will be someone else's story.

Acknowledgments

First I would like to dedicate this monograph to my dear friend and colleague Soren Buus. Thanks to Soren for the many one-on-one, engineer-to-engineer conversations about psychophysics and his significant contribution to my understanding of this field and for his many comments on an early version of this monograph. Soren sorely is missed.

Thanks to Leo Beranek, Arthur Boothroyd, Andreas Buja, Mark Hasagawa-Johnson, Steve Levinson, Harry Levitt, George Miller, Richard Sproat, and Vinay Vaishampayan for *many* helpful discussions.

I would also like to especially thank Terry Chisolm, and Tina Rankovic for extensive comments on an early drafts of the manuscript, and finally Zoran Cvetković for months of interaction and input, and providing me with many extensive revisions in the later months of its preparation. Finally, thanks to David Pinson and Robert Remez, who first asked me to write this document, as a book Chapter, and then excluding me when it became to long, thereby allowing it (forcing it) to become this monograph. It turned out that my attempt toward writing the book chapter ultimately stretched well beyond its intended scope. The lesson is, don't submit an 75 page document when 25 were requested. This is a painful lesson that I own.

CHAPTER 1

Introduction

As discussed by George Miller in *Language and Communication* (Miller, 1951), speech intelligibility depends on a code that is incredibly robust to mangling distortions. Can we analyze this code using the scientific method? Miller's analysis provides a method that is both scientific and artful. His scientific analysis critically depends on the use of statistics, information theory, and psychophysical methods.

The science of intelligibility is a science of the error analysis of human speech communication. The goal in science is always to make a mathematical model. Data is collected and tested against the model. This chapter is a review of what is known about modeling *human speech recognition* (HSR) (see Table 1.1 for each abbreviation). A model is proposed, and data are tested against the model.

Why would one wish to study human speech perception? First a great deal of work has been done on this problem over the past 80 years. The time is now ripe to review this large literature. There seem to be a large number of theories, or points of view, on how human speech recognition functions, yet few of these theories are either quantitative or comprehensive. What is needed is a set of models that are supported by experimental observation and that quantitatively characterize how human speech recognition really works. Finally there is the practical problem of building a machine recognizer. One way to do this is to build a machine recognizer based on the reversed engineering of human recognition. This has not been the traditional approach to automatic speech recognition (ASR).

As we shall see, human recognition performance drops to chance levels at a wideband SNR somewhere between -20 and -25 dB and saturates at a minimum error rate when the SNR is between -6 and 0 dB. The actual numbers depend on the complexity (i.e., the entropy) of the recognition task. Machine performance starts degrading by $+20$ dB SNR and approaches chance level near 0 dB SNR. This large discrepancy in performance is called the *robustness* problem, which refers to the sensitivity of the score to the SNR. What is needed is some insight into why this large difference between human performance and present day machine performance exists. I believe we can answer this question, and this is one of the goals of this monograph.

In many of the early studies of human speech recognition, many language *context* effects were controlled by testing with *nonsense syllables*. When listening to meaningful words and sentences, people report what they understand, leaving many errors they hear unreported. When listening to nonsense speech, having limited language context constraints, people report what they actually hear. Thus to meaningfully study the decoding of speech sounds, one must carefully control for meaning (i.e., context). One does this by controlling the *context channel* with the help of a maximum entropy (MaxEnt) source, more commonly called nonsense syllables.

The channel, a mathematical correlate of a pair of wires, with additive noise and codecs attached, is a key concept devised by Shannon in his *Theory of Information* (Shannon, 1948). As developed in Miller's book, information and communication theory form the underlying scientific basis for understanding the speech code. The channel and the entropy measure of the source are fundamental building blocks of any theory of communication. Speech and language are no exception.

The term *nonsense syllables* is a serious misnomer since a list of such syllables contains meaningful words. What is required from such lists is that the probability of each phone (i.e., phoneme) has equal probability. A proper name for such sounds is *maximum entropy syllables*. For example, if the syllables are of the form consonant-vowel (CV), VC and CVC, and each of the Cs and Vs are chosen from a fixed alphabet in an equally likely way, then this corpus of *MaxEnt Syllables* contains all possible meaningful words of 2 and 3 phone duration. The use of such MaxEnt syllables is required by Shannon's model of communication, as captured by the source-channel model.

There seems to be some widely held beliefs that are not supported by the science. One of these is that improving the language processing in automatic speech recognition will solve the robustness problem. Human performance starts with the detection of basic speech sounds. The data show that when the speech is masked by noise, and is at the detection threshold, humans are able to categorize the sounds into broad phonetic categories. This detection threshold lies below -18 dB wideband SNR, where classification is well above chance. Language models, which are modeled by context processing (i.e., Markov chains and formal grammars), cannot play a role in recognition until the error rate is less than some critical number. One could argue about what this critical value is, but a reasonable number must be near a 50% raw phone error. Until this critical value is reached, language (and therefore any language model) does not have enough input information to significantly improve the score.

As the score reaches 100%, language (context) plays a key role (context is best at fixing sparse, complex [i.e., low entropy] errors). However, when studying the robustness problem, one must carefully control for context and work at high-error rates. Thus language, while critically important, plays a role only when the score is above this critical threshold, and plays no role in understanding the robustness problem. In fact, when it is not controlled, context gets in the way of understanding robustness and *becomes* the problem.

This myth, in the power of context at low SNRs, is further complicated by the fact that human language processing performance (phonology, phonotactics, syntax, morphology, prosody, etc.) far exceeds the performance of today's machine language processing. This leads to the following dilemma for ASR, as predicted in 1963 by Miller and Isard: Both ASR's front-end phone error rate *and* its back-end context processing are significantly worse than those of HSR. Language models can never achieve the desired goals of solving the robustness problem because it is the front end that accounts for the errors causing the robustness issues. Thus we must deal directly with the front-end problems of talker variation, noise, and spectral modifications, *independent* of context effects as HSR does. This view is not to depreciate the importance of context, rather it is an attempt to clarify the statement that improved context processing cannot solve the noise robustness problem.

Another myth is that human listeners cannot hear subtle distinctions of production that are not phonemic in their native language. It is an observable fact that when we speak in everyday conversations, we mispronounce speech sounds a large percentage of the time. The severity of these errors is a function of the speaker and the conditions. For example, a person with a strong accent is an individual with a bias for "mispronouncing" the sounds. This bias has many sources. The talker's first language is the most important factor. A well known example is the L/R discrimination in Japanese talkers, when speaking English. Initially it was believed that one cannot hear (and therefore cannot learn to correct) these biases. However it is now believed that one may reduce, and in rare cases even eliminate these errors, given proper training.

Yet another myth is the importance of *coarticulation*. It is well known that the formants for a given vowel are greatly affected by the consonant preceding or following the vowel. This may be seen in the spectrogram. Coarticulation is frequently identified as a difficult, or even the key, problem, at the root of vowel classification, because the variability of the formants due to this coarticulation effect renders the utility of the formant frequencies unreliable as a feature set.

The circular nature of this argument is stunning. It starts by assuming that the information is in the formant frequencies, and then when these features turn out to be unreliable, summarizes the dilemma as coarticulation. Let us suppose that the vowel speech code is not simply the formant frequencies. Would we come to the same conclusion in that case? In my view, no. When one looks for psychophysical evidence of coarticulation, there are none. Fletcher (1921) was the first to clearly show this with his models of MaxEnt syllable perception, which showed that the probability for correctly identifying MaxEnt syllables may be treated as an independent product of the component phone probabilities. These models were later confirmed by Boothroyd (1968), and many others.

Unfortunately, these many myths die very slowly.

The art of studying speech intelligibility is to restrict the problem in such a way that progress may be made. Intelligibility depends on *visual* as well as auditory cues. In this chapter our discussion will be limited to that subset of intelligibility which is unique to acoustic speech, *without visual input*. Thus we restrict the model to the *auditory channel*, in the absence of the *visual channel*.

A tool critical to the study of human language perception is the use of MaxEnt (i.e., nonsense) syllables. Such sounds were first extensively used in the Bell Labs studies to control for language context effects. We will model context as a channel, which carries the side information and is helpful to the listener.

While the use of MaxEnt syllables reduces the context effect, the complete elimination of all context channels is an impossible task, and even undesirable. English MaxEnt sounds are a distinct subset of speech sound. Tonal sounds are common in many languages such as Chinese, but are absent in English. Thus the subset of English MaxEnt sounds, while rich enough to encode English, is a distinct subset of human vocalizations. The best we can do is attempt to characterize these more subtle context channels, not eliminate them, as we attempt to saddle this untamable "context beast."

Intelligibility is the identification of meaningful speech, while *articulation* is the identification of *MaxEnt* speech sounds (see Table 1.2 for important definitions). Understanding intelligibility scores requires models of phonology (phone interactions), lexicality (vocabulary), grammar (form), and semantics (meaning). To understand articulation scores, only models of phonetics are required, by design, by carefully controlling for as many of these context channels as possible.

The word articulation is a tricky term, as it has strong meanings in both the production (physical) and perceptual (psychophysical) domains. An *articulatory feature* is a speech production concept, whereas the *articulation score* is a perceptual concept. It is quite unfortunate, and very confusing, that this word has these two different but related meanings. We have inherited these terms from the long past, and thus must deal with this confusion. One way to deal with this problem of terminology would be to create new terms. However this would just obscure and confuse the situation further, so I have avoided that approach. It is better to be aware of, understand, and carefully parse the two meanings.

Is there any information left in speech after these major information channels, visual and context, have been removed? Emphatically, *yes!* In fact humans do quite well in identifying basic speech sounds well below 0 dB SNR, without these powerful information side channels. The reasons for this *natural robustness* in HSR are becoming increasingly clear, and are the topic of this monograph.

1.1 PROBLEM STATEMENT

Articulation has been studied since the turn of the twentieth century by teams of physicists, mathematicians, and engineers at Western Electric Engineering, and later at The Bell Laboratories. A key science, information theory, also developed at Bell Laboratories, is frequently emphasized, yet rarely applied to the field of speech recognition. As a result, the source of robustness in HSR is poorly understood. For example, many believe that robustness follows from context.

An example of this view may be found in Flanagan's classic text *Speech Analysis Synthesis and Perception* (Flanagan, 1965, p. 238)

> Items such as syllables, words, phrases, and sometimes even sentences, may therefore have a perceptual unit. In such an event, efforts to explain perception in terms of sequential identification of smaller segments would not be successful.

This view is not supported by the data, as we shall see. Speech is first detected in white (uniform spectrum) masking noise at about −25 dB SNR, and basic sound classes are easily discriminated at −20 dB SNR. Language (sentences, phrases, and words), which depend on context, cannot play a role until a sufficient fraction of the sounds have been decoded.

Only data and experiment can resolve the fundamental question: "What are the smallest units that make up the building blocks of oral to aural speech perception?" The resolution of this question may be answered by modeling intelligibility and articulation data. The robustness of MaxEnt speech has been measured with a confusion matrix based on MaxEnt speech sounds (Campbell, 1910; Miller and Nicely, 1955), which is denoted the *articulation matrix* (AM). Many important issues regarding AM data remain unstudied.

This brings us to Miller's unsolved problem, the decoding of MaxEnt speech sounds, which have been mangled by filtering and noise. What are the remaining information channels that need to be accounted for, and how can we model them? This monograph will explore this question in some detail. We begin in Section 1.2 with some definitions and an overview of the robustness problem. We then proceed to Chapter 2 with a literature review of articulation testing and the articulation index (AI), and the work of George Miller, who first controlled for the *articulation test entropy* $\mathcal{H}$ with the use of close set testing. This leads us to an AI analysis of Miller and Nicely's consonant confusion data. In Chapter 3 we look at the nature of the context channel with a review of the context models of Boothroyd and Bronkhorst. From this higher ground we model how aural speech is coded and processed by the auditory system.

1.2 BASIC DEFINITIONS AND ABBREVIATIONS

Tables 1.1 and 1.2 provide key abbreviations and definitions used throughout the paper. While it is important to carefully define all the terms, this section could be a distraction to the flow of the discussion. Rather than bury these definitions in an appendix, I have placed them here, but with the warning that the reader should not get bogged down with the definitions. I suggest you first skim over this section, to familiarize yourself with its content. Then proceed with the remainder of the material, coming back when an important idea or definition is unclear. Refer to

TABLE 1.1 Table of abbreviations

ASR	automatic speech recognition
HSR	human speech recognition
CV	consonant–vowel (e.g., "pa, at, be, c")
CVC	consonant–vowel–consonant (e.g., "cat, poz, hup")
snr	signal-to-noise ratio (linear units) Eq. (2.15)
SNR	$20 \log_{10}(snr)$ (dB units)
AI	articulation index
AI_k	specific AI (dB/dB units) Eq. (2.16)
PI	performance intensity function $P_c(SNR)$
AM	articulation matrix
VOT	voice onset time
ZP	zero predictability ($-$ semantics; $-$ grammar)
LP	low predictability ($-$ semantics; $+$ grammar)
HP	high predictability ($+$ semantics; $+$ grammar)
ERP	event related (scalp) potential

the tables as a quick guide, and then the text, once the basic ideas of the model are established.

It is essential to understand the definition of *articulation*, the *event*, and why the term *phone* is used rather than the popular term *phoneme*. A qualitative understand of *entropy* is also required. All of the required terms are now carefully defined.

The phone vs. phoneme: The *phone* is any basic speech sound, such as a consonant or vowel. It must be carefully distinguished from the *phoneme*, which is difficult to define because every definition incorporates some form of *minimal meaning*. A definition (see Table 1.2) has been chosen that is common, but is not universally agreed upon.

TABLE 1.2 Definitions

TERM	DEFINITION
phone	A consonant (C) or vowel (V) speech sound
syllable	A sequence of C's and V's, denoted {C,V}
word	A *meaningful* syllable
phoneme	Any equivalent set of phones which leave a word meaning invariant
allophones	All the phone variants for a given phoneme
recognition	Probability measure P_c of correct phone identification
articulation	Recognition of nonsense syllables MaxEnt ({C,V})
intelligibility	Recognition of words (i.e., meaningful speech)
confusion matrix	Table of identification frequencies $C_{sb} \equiv C_{b\|s}$
articulation matrix	A *confusion matrix* based on nonsense sounds
robustness	Ratio of the conditional entropies for two conditions to be compared
event	A perceptual feature; multiple events define a phone
trial	A single presentation of a set of events
state	A values of a set of events at some instant of time
state machine	A machine (program) that transforms from one state to another
noiseless state machine	A deterministic state machine
p_n	Probability of event n, of N possible events
information density	$I_n \equiv \log_2(1/p_n), \quad n = 1, \dots, N$
entropy	Average information: $\mathcal{H} \equiv \Sigma_{n=1}^{N} p_n I_n$
conditional entropy	A measure of context: high entropy $\Longrightarrow$ low context
context	Coordinated combinations of events within a trial
message	Specific information transmitted by a trial (e.g., a syllable)

I shall argue that meaning is irrelevant to the speech robustness problem. During World War II, people were trained to transcribe languages that they did not understand, and they did this with agility and fidelity. Fletcher AI theory (1921–1950) was based on MaxEnt CV, VC, and CVC syllables (i.e., maximum entropy syllables). Shannon formulated Information Theory based on entropy measures. Miller and Nicely's classic study used isolated consonants, which by themselves have no meaning (Miller and Nicely, 1955).

Thus one may proceed with the study of human speech recognition, without the concept of meaning, and therefore the phoneme. This view has a nice parallel with Shannon's theory of information, which specifically rejected meaning as relevant (Shannon, 1948).

It is difficult to argue strongly for the importance of the phoneme, whose definition depends on meaning, if meaning plays little or no role in peripheral language processing (the robust identification of unit phones).

A *syllables* is one or more phones. A *word* is a syllable with meaning (it is found in a dictionary).

Recognition is the probability of correct average identification, denoted P_c. *Recognition error*, as given by

$$E\% \equiv 100(1 - P_c),$$

is typically quoted in percent,[1] and is the sum over all the individual sound confusions. The recognition P_c (and thus the corresponding recognition error $E\%$) is a function of SNR. When the recognition is measured as a function of the signal to noise ratio, it is frequently called a *performance-intensity* (PI) function, as a function of the SNR in dB, denoted P_c (SNR).

Articulation is the recognition of MaxEnt speech sounds (e.g., nonwords), while intelligibility is the recognition of *meaningful* sounds (dictionary words) (Fletcher, 1929, p. 255).

---

[1] The symbol $\equiv$ is read "equivalence" and means that the quantity on the left is defined by the quantity on the right.

The *articulation matrix*, denoted $P_{h|s}(\text{SNR})$, tabulates the recognition scores of hearing sound h after speaking sound s, where h and s are integers.

Robustness: An important, but again difficult, concept to define is that of *robustness* (an important topic of this chapter). The first property of robustness must be that it is a relative measure. Second we would like a measure that is defined in terms of bits. Specific examples of the use of such a relative measure can help us to further nail down the full definition: An important example is the robustness of one sound versus another (i.e., /pa/ vs. /ma/). Another example is the robustness of one group of sounds, say the nasals, against another group of sounds, say the fricatives.

In each example there are two cases we wish to compare, and we would like a measure that tells us which is more robust. The candidate measure for the first example of two sounds is the conditional entropy, defined as

$$\mathcal{H}(h \mid s) = -\sum_h P_h \, \log_2(P_{h|s}),$$

which is just the entropy of row s of the articulation matrix $P_{s,h}$.

This measure is in bits, as required, for each spoken sound s. This measure has the unfortunate property that it becomes smaller as the sound becomes more certain, which is backward from a robustness measure. If we define the relative measure as the ratio of two conditional entropies, for the two different sounds, then we have a measure that increases as the score increases. For example, the robustness of spoken sound s_2 relative to that of s_1 would be

$$\mathcal{R}(s_2/s_1) = \frac{\sum_h P_h \, \log_2(P_{h|s_1})}{\sum_h P_h \, \log_2(P_{h|s_2})}.$$

This measure would increase if s_2 is more robust (has a smaller conditional entropy) than s_1.

As a second example lets take the robustness of intelligibility vs. articulation (i.e., the effect of context). In this case the robustness due to intelligibility would be taken to be

$$\mathcal{R}(\mathcal{I}/\mathcal{A}) = \frac{\sum_{s,h} P_{s,h}(\mathcal{A}) \log_2(P_{h|s}(\mathcal{A}))}{\sum_{s,h} P_{s,h}(\mathcal{I}) \log_2(P_{h|s}(\mathcal{I}))},$$

where $P(\mathcal{I})$ is with context (intelligibility) and $P(\mathcal{A})$ is with no context (articulation).

If we wish to compare the robustness of ASR and HSR, the robustness would then be

$$\mathcal{R}(\text{HSR}/\text{ASR}) = \frac{\sum_{s,h} P_{s,h}(\text{ASR}) \log_2(P_{h|s}(\text{ASR}))}{\sum_{s,h} P_{s,h}(\text{HSR}) \log_2(P_{h|s}(\text{HSR}))}.$$

For these last examples it makes sense to restrict comparisons to cases that have the same maximum entropy, namely for which the corpus is the same size, if not identical and the same SNR.

Events: In the speech perception literature the terms *articulatory feature, perceptual feature*, and *distinctive feature* are commonly used, even interchangeably (Parker, 1977). For example, *voicing, nasality*, and the *place* of a constriction in the vocal tract, which occurs when forming a speech sound, constitute typical articulatory features. The term voicing is frequently spoken of as both a physical and a perceptual feature. It seems wise to choose a new word to represent the *perceptual correlates* of speech features. We use the word *event* to deal with such meaning.

The basic model of psychophysics and of the observer is shown in Fig 1.1. As for the case of intensity and loudness, we need a language for relating *perceptual features* (Ψ variables) to physical *articulatory features* (Φ variables). Thus we speak of the *event* when referring to the Ψ correlate of an speech Φ feature. For example, it might turn out that the Ψ-event corresponding to the Φ-feature *voicing* is determined by quantizing the so-called Φ voice onset time (VOT) to some fixed time range of values. A Φ-VOT between 0 and 30 ms might be Ψ-VOICED, while Φ-VOTs greater than 30 ms might be Ψ-UNVOICED.

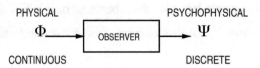

FIGURE 1.1 : The basic model of an observer with the physical variables Φ on the left and the psychophysical variables Ψ on the right. An example is acoustic intensity (the Φ or physical intensity) and loudness (the Ψ or psychoacoustic intensity). In the case of speech perception we treat physical variables as analog (continuous) and psychophysical variables as discrete, as in the case of events.

The *event* must be measured by experimental outcomes, expressed as a probability, rather than assumed *a priori*, as in the case of *distinctive features*. The articulation matrix $\mathcal{A}(SNR)$ is the measure of these experimental outcomes. We do an experiment where we repeat the stimulus many times, and we then define a probability measure of the underlying binary *event*, in terms of the frequency of its observation, based on a large number of subjects and a large number of talkers.

Each presentation and reception is called a *trial*. This idea is a formal one, as described by books on communication theory (Wozencraft and Jacobs, 1965, Ch. 2) and probability theory (Papoulis, 1965, Section 2.2). These definitions, of a *trial* and an *event*, as defined in this mathematical literature, are ideally suited to our purpose.

When groups of events are mathematically bound together at an instant of time, the group is called the *state* of the system. As an example, think of the events that define the state of a phone. A machine state (think computer program) is typically pictured as a box that transforms an input state into an output state. When the state machine is deterministic, it is called a *noiseless state machine*. During training (the learning phase), the state is not deterministic, but such a learning mode is considered to be an exception for the purpose of modeling the state machine. We view the auditory brain as a state machine decoding the events coming out of many event processors, having inputs from the cochlea. This model structure represents the "front end" of the HSR system.

SNR: The SNR plays a very important role in the theory of HSR because it is the underlying variable in the articulation index measure. The detection of any signal is ultimately limited by detector noise. This leads to the concept of an internal noise, specified as a function of frequency. It is the *internal* signal-to-noise ratio $SNR(f)$, a Ψ variable, that ultimately determines our perceptual performance (French and Steinberg, 1947; Allen and Neely, 1997). This quantity must be inferred from external measurements.

An example is instructive: The external SNR of a pure tone, in wide band noise, is not perceptually meaningful since a relevant noise bandwidth must be used when calculating the detection threshold. This bandwidth, called the *critical bandwidth*,[2] is cochlear in origin, since the internal $SNR(f)$ depends on cochlear filtering. The discovery of the cochlear critical bandwidth marked the recognition of this fact (Fletcher and Munson, 1937; Fletcher, 1938; French and Steinberg, 1947; Allen, 2001).

Even though speech is a wide band signal, exactly the same principle applies to the detection of speech. The detection threshold for speech sounds are determined by the same cochlear critical bandwidth. Unlike the tonal case, the peak to RMS ratio of the speech in a critical band becomes a key factor when estimating the speech detection threshold.

These basic issues of speech detection and articulation were well understood by Fletcher and his colleagues Wegel, Steinberg, and Munson, and were repeatedly described in many of their early papers. These points will be carefully reviewed in the next chapter.

Two different notations for the signal-to-noise ratio shall be used: $snr \equiv \sigma_s/\sigma_n$ and $SNR \equiv 10\log_{10}(\sigma_s^2/\sigma_n^2)$. Each of these measures will be indexed by integer k to indicate each critical band. Thus in critical band k, $snr_k \equiv \sigma_{s,k}/\sigma_{n,k}$.

---

[2] In many of the early papers the level of a tone in noise above threshold, expressed in dB-SL, was commonly denoted by the variable Z (French and Steinberg, 1947, Eq. 2, p. 97). This definition explicitly accounts for the critical bandwidth of the ear.

Context and entropy: The concept of *context* in language is ubiquitous. Context results from a time-correlated sequence of speech units, leading to the higher probability of predicting a word, based on the preceding words. Mathematically this can be expressed as

$$P_c(x_n) < P_c(x_n \,|\, C = x_1 x_2 \cdots x_{n-1}), \tag{1.1}$$

where x_n are speech units and C is the conditioning context based on the last $n-1$ sound units (i.e., words or phonemes). If x_n are random unrelated units, then the sequence $x_1, x_2, \ldots, x_{n-1}$ does not change the score of x_n; i.e., the conditional recognition of x_n is the same as that of the isolated speech unit.

It is critically important to control for context effects when studying speech recognition. Real words have greater context than randomly ordered meaningless speech sounds, which ideally, would have none. Meaningful HP sentences have greater context than MaxEnt ZP sentences. One classic way of modeling context is with Markov models (Shannon, 1948, 1951).

By *redundancy* we mean the repetition of events within a trial.[3] Sometimes redundancy depends on context, as in the example *Sierra Mountains*.[4]

The *information density* I_n is defined as the log base 2 of the reciprocal probability p_n. The log base 2 is a simple transformation that gives units of bits. The important concept here is reciprocal probability, so that a rare event (small probability) is defined as having large information. The concept of probability, and thus of information, requires a *set of outcomes*. The vector of probabilities $[p_n]$ requires an index n labeling N possible outcomes, while element p_n measures the relative frequency (parts of a whole) of these outcomes, which obey the condition $\sum_n p_n = 1$.

Entropy is the average amount of information, as computed by taking a weighted average of the information density, as shown in Table 1.2. When all the

---

[3] This term has been mathematically defined by Shannon in his classic paper (Shannon, 1948, p. 24).

[4] The word Sierra means mountain in Spanish (a language context).

FIGURE 1.2 : Who is this monkey (chimp in this case) thinking of, and what does she want to say? How many messages can a monkey type? This picture epitomizes the concept of entropy. It is highly unlikely, yet not impossible, that through random typing, this monkey will produce a work of Shakespeare, corresponding to an astronomically small entropy. Of course even a single sentence of Shakespeare is virtually impossible for the monkey to produce.

outcomes are equal (i.e., $p_n = 1/N$) the entropy $\mathcal{H}$ is maximum, and the information is minimum (Cover and Thomas, 1991). Fig. 1.2 is an epitome of entropy. How many monkeys would it take to produce a work of Shakespeare? The entropy of such a document is very low. The number of monkeys working in parallel that are needed to have one of them produce such a document is astronomical.

1.3 MODELING HSR

It is important to develop a model of human speech recognition (HSR) that summarizes what we know in a succinct manner. A model is presented in Fig 1.3, which shows the structural relations between the various quantitative probabilistic measures of recognition.

It is widely accepted, typically with no justification, that HSR is modeled by a *front end* driving a *back end*. These two terms have been used loosely in the past, and they can have different meanings in different fields (e.g., speech, psychology, and physiology). We shall define the front end as the acoustics processing and *event extraction* stage and the back end as the *context processing* stage. An excellent quantitative justification for doing this is provided by the work of Boothroyd (1968) and later Bronkhorst (Bronkhorst *et al.* 1993), who defined a front end and a back end in a mathematical model of context processing. In 1968 Boothroyd modeled the effects of word recognition, given phone scores, as a contextual constraint, and made empirical models to account for this context effect (Boothroyd, 1968, 1993; Boothroyd and Nittrouer, 1988). Bronkhorst *et al.* (1993) integrated Fletcher's AI model and generalized Boothroyd's context models to include all possible combinations of recognition errors, thereby quantitatively extending context models. They also derived model weighting coefficients from first principles, using a lexicon.

In the model shown in Fig 1.3, all of the recognition errors in HSR are a result of event extraction labeling errors, as depicted by the second box and modeled by the articulation-band errors e_k. In other words, *sound recognition errors are modeled as a noise in the event conversion from analog to discrete "objects."* I will argue that much of this frontend event processing is implemented as *parallel processing,*[5] which is equivalent to assuming that the recognition of events is independent, on average, across cochlear frequency bands.

As shown in Fig 1.3, the input speech signal is continuous, while the output stream is discrete. Somewhere within the auditory brain discrete decisions must be made. A critical aspect of our understanding is to identify at what point and at what level this conversion from continuous to discrete takes place. I will argue that this conversion is early, at the event level. Once these decisions have been made,

---

[5] The idea behind *parallel processing* will be properly defined in Section 2.3.1.

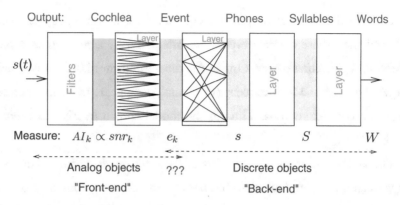

FIGURE 1.3 : Model block diagram summary of speech recognition by humans. At the top of each block is a label that attempts to identify the physical operation, or a unit being recognized. The labels below the boxes indicate the probability measure defined at that level. See the text for the discussion of objects, at the very bottom. The speech $s(t)$ enters on the left and is processed by the cochlea (first block), breaking the signal into a filtered continuum of band-passed responses. The output of the cochlea is characterized by the specific AI_k, a normalized *SNR*, expressed in dB units. The second box represents the work of the early auditory brain, which is responsible for the identification of events in the speech signal, such as onset transients and the detection of basic measures. The third block puts these basic features together defining phones. The remaining blocks account for context processing.

the processing is modeled as a *noiseless state machine* (i.e., a state machine having no stochastic elements).

When testing either HSR or ASR systems, *it is critical to control for language context effects*. This was one of the first lessons learned by Fletcher *et al.* that context is a powerful effect, since the score is strongly affected by context.

The HSR model of Fig 1.3 is a "bottom–up," divide and conquer strategy. Humans recognize speech based on a hierarchy of context layers. Humans have an intrinsic robustness to noise and filtering. In fact, the experimental evidence suggests that this *robustness* does not seem to interact with semantic context (language), as reflected by the absence of feedback in the model block diagram.

Their is a long-standing unanswered question: *Is there feedback from the back end to the front end?* The HSR model shown in Fig 1.3 assumes that *events* are extracted from the cochlear output in frequency regions (up to, say, the *auditory cortex*), and then these discrete events are integrated by a *noiseless state machine* representing the cerebral cortex. One of the most important issues developed here is that *front-end* phone feature recognition analysis appears to be independent of the *back-end* context analysis. Thus in the model shown in Fig 1.3 there is no feedback.

The auditory system has many parallels to vision. In vision, features, such as edges in an image, are first extracted because in vision, entropy decreases as we integrate the features and place them in layers of context. This view is summarized in Fig 1.3 as a *feed-forward* process. We recognize events, phones, phonemes, and perhaps even words, without access to high-level language context. For designers of ASR systems, this is important and good news because of its simplicity.

As early as 1963 Miller and Isard made a strong case against the use of Markov models in speech recognition, using an argument based on robustness, in an apparent reaction to the use of language (context) models (i.e., in ASR applications this amounts to hidden Markov models, or HMM) for solving the robustness problem. While language context is key in reducing many types of errors, for both ASR and HSR, the front-end robustness problem remains. Although it is widely believed that there is much room for improvement in such language models (Miller, 2001), it now seems clear that even major context processing improvements will not solve the ASR noise robustness problem. We know this from an analysis of data from the literature, which shows that humans attain their inherent robustness to background noise early, independent of and *before* language context effects.

Nonsense speech sounds may be detected starting at about −20 dB SNR (wideband). Context effects begin having an impact when the score is above 50% and have a minimal effect at low articulation scores below this score.

As discussed in Section 1.3.1, randomizing the word order of grammatically correct sentences degrades the SNR by 6–10 dB. Miller argues that such a word-randomizing transformation would have much larger performance degradation on a Markov driven ASR system, which systematically depends on word order (see Section 1.3.1).

1.3.1   Context Models

An example of a context effect: A detailed example of the utility of context in HSR was demonstrated by Miller (1962). This example stands out because of the early use of ideas from information theory to control for the entropy of the source, with the goal of modulating human performance via context. The experiment was simple, yet it provides an insight into the workings of context in HSR.

In this experiment 5 groups of 5 words each make up the test set. This is a *closed-set*[6] listening task with the number of words and the SNR varied. There are four conditions. For test condition one the subjects are shown 1 of the 5 lists, and they hear a word from that list. For the other three conditions the subjects are shown 1 list of all the 25 words. The probability correct $P_c(SNR)$ was measured for each of the four conditions:

- 5 words;

- 5 word grammatically correct sentences, chosen from the 25 words;

- 25 words;

- nongrammatical sentences chosen from the 25 words.

As described in Fig. 1.4, in condition (1) 5 word lists are used in each block of trials. The lists are randomized. The subject hears 1 of 5 words, degraded by noise, and is asked to pick the word from the list. In condition (3) the number of words is increased from 5 to 25, causing a reduction of 4 dB in performance (at the 50% level). These two conditions (1 and 3) were previously studied in a classic paper (Miller *et al.*, 1951), which observed that the size of the set of CVCs has a large impact on the score, namely $P_c(SNR, \mathcal{H})$ depends on the entropy $\mathcal{H}$

---

[6] A *closed-set* test is one with a limited number of outcomes that are known *a priori* to the subjects.

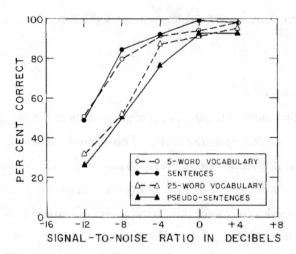

FIGURE 1.4 : This figure, from Miller (1962), summarizes the results of a four-way experiment, performed as a function of the SNR. Test 1 (open circles, dashed line) shows $P_c(SNR)$ for 5 word vocabularies, with no context. In test 2 (closed circles, solid line) 5-word sentences were made from the 5, 5-word lists. As an example "Don brought his black socks." The word "Don" was one of the 5 possibilities [Don, He, Red, Slim, Who]. For tests 1 and 2, $P_c(SNR)$ is the same. Test 3 (open triangles, dashed line) was to test using the larger corpus of one of the 25 words, spoken in isolation. Test 4 (closed triangles, solid line) was to generate "pseudo-sentences" by reversing the order of the sentences of test 3. Going from 5 to 25 isolated words (test 1–3) causes a 4 dB SNR reduction in performance at the 50% correct level. Presenting the 25 words as pseudo-sentences, that make no sense (test 4), has no effect on $P_c(SNR)$. However, adding a grammar (test 2) to a 25 word test returns the score to the 5 word test. In summary, increasing the test size from 5 to 25 words reduces performance by 4 dB. Making 5 word grammatically correct sentences out of the 25 words restores the performance to the 5 word low entropy case.

of the task. In condition (2), the effect of a grammar context is measured. By placing the 25 words in a context having a grammar, the scores returned to the 5 isolated word level (condition 1). When sentences having no grammar (pseudosentences) were used (condition 4), generated by reversing the meaningful sentences of condition (2), the score remains equal to the 25 isolated word case of condition (3).

Thus the grammar in experiment (2) improves the score to the isolated word level (1), but not beyond. It probably does this by providing an improved framework for remembering the words. Without the grammatical framework, the subjects become confused and treat the pseudosentences as 25 random words (Miller and Isard, 1963).

1.4   OUTLINE

The monograph is organized as follows: Sections 2.1 and 2.2 summarize important results from the 30 years of work (1921–1950) by Fletcher and his colleagues, which resulted in *articulation index theory*, a widely recognized method of characterizing the information bearing frequency regions of speech. We shall show that the AI is similar to a *channel capacity*, which is a key concept from information theory defining the maximum amount of information that may be transmitted on a channel. Section 2.4 summarizes the speech work of George Miller. Miller showed the importance of source entropy (randomness) in speech perception. He did this by controlling for both the cardinality (size of the test corpus), the signal to noise ratio of the speech samples and the context. Section 2.7 discusses the validation and section 2.8 criticisms of articulation index theory. Section 3 discusses the importance of context on recognition, summarizing key results. For continuity, research results are presented in chronological order.

CHAPTER 2

Articulation

In 1908 Lord Rayleigh reported on his speech perception studies using the "Acousticon LT," a commercial electronic sound system produced in 1905. As shown in Fig. 2.1,[1] it consisted of a microphone and four loudspeakers, and was sold as a "PA" system. Rayleigh was well aware of the importance of the bandwidth and blind-speech-testing to speech perception testing (Rayleigh, 1908). Apparently Rayleigh was the first to use "electro-acoustics" in speech testing.

Rayleigh's 1908 work was extended by George A. Campbell (Campbell, 1910), a mathematician by training (and an amazing engineer, in practice), trained at Harvard and MIT, and employed by AT&T to design the transmission network. Campbell is well known for his invention of the hybrid network (a 2–4 wire transformer), the loading coil (which greatly extended the frequency response of the telephone line), and perhaps most important, the "wave" filter in 1917 (Campbell, 1922, 1937; Van Valkenburg, 1964; Wozencraft and Jacobs, 1965, p. 1).[2]

It was the development of the telephone (circa 1875) that both allowed and pushed mathematicians and physicists to develop the science of speech perception. Critical to this development was probability theory. One of the main tools was the *confusion matrix* $P_c(h_i \mid s_j)$, which estimates the probability P_c of hearing speech sound h_i when speaking sound s_j (Campbell, 1910; Fletcher, 1929, pp. 261–262).

---

[1] http://dept.kent.edu/hearingaidmuseum/AcousticonLTImage.html.

[2] Apparently these filters were first invented to filter speech so as to aid in intelligibility testing of telephone channels.

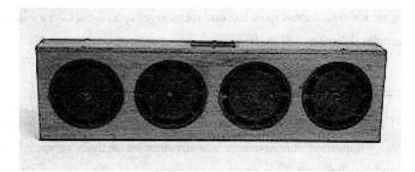

FIGURE 2.1: The Acousticon LT was invented in about 1905.

From 1910 to 1950 speech perception was extensively studied by telephone research departments throughout the world. However, it was the work of Harvey Fletcher in 1921 that made the first major breakthroughs (Fletcher, 1921). By 1930 millions of dollars were being spent each year on speech perception research at the newly created Bell Labs (Rankovic and Allen, 2000). The key was his quantification of the transmission of information, as characterized by MaxEnt syllable error patterns. Fletcher's final theory was not published until 1950, following his AT&T retirement (Fletcher and Galt, 1950). A review of Fletcher's work may be found in (Allen, 1994, 1996).

The next breakthroughs were provided by George Miller and his colleagues, working at Harvard's Cruft Acoustics Lab during and following World War II. Miller used concepts from information theory, developed at Bell Labs, by Claude Shannon (Shannon, 1948), to quantify speech entropy. In 1951 Miller *et al.* quantified the effect of entropy on recognition by studying the intelligibility as a function of the number of choices. The insight for these studies came from his work on speech masking, where AI theory played an important role (Miller, 1947a). The same year (1951) Miller published his treatise on speech and information theory, which was the first to explore the application of Shannon's theory to HSR (Miller, 1951). In 1955 Miller and Nicely published the first extensive confusion matrices (i.e. AM) for filtered and noisy CV sounds.[3] Starting with the articulation measurements by

---

[3] Earlier but limited AMs were published by Campbell (1910) and Knudsen (1929).

Fletcher, we know that most speech sound features are spread out over a wide band of frequencies (Fletcher, 1922a, 1922b).

2.1   FLETCHER AND GALT (1950)

Harvey Fletcher had an amazing career, and lived to 96 years (Allen, 1994). One of his many important contributions was his detailed analysis of speech articulation and intelligibility. This work is summarized by *articulation index theory*, based on the idea of independent recognition of speech events in frequency bands (Fletcher, 1921; French and Steinberg, 1947; Fletcher and Galt, 1950).

It may not be widely appreciated that his demonstration of band-independence (the basic premise of the AI) was not assumed, but was *deduced* from an additivity principle, and demonstrated by experimental results. His idea was to find a functional transformation on band articulations that would make the transformed articulations additive. The data-collection phase took tens of years, and costed millions of dollars. As a result of this massive effort, the United States had the highest quality telephone service on the planet.

The telephone began as a very crude device, initially providing poor intelligibility. It was immediately clear that dramatic improvements were necessary if they were going to build and maintain a national telephone network. Fortunately there was great economic incentive to do so. The tedious speech testing work easily paid its own way. After years of experimentation with speech sounds, Fletcher and his team developed an efficient speech test. The key was to use nonsense speech sounds composed of a balance of CVC, CV, and VC sounds. The exact balance was determined by listening to live conversations on the network. At the time (unlike today) the scientific observation of telephone calls was not illegal. J.Q. Stewart played an important role in developing the application of probability theory to speech (Fletcher and Galt, 1950; Rankovic and Allen, 2000; Allen, 1994).

The experiment: Articulation testing consists of playing nonsense syllables composed of 60% CVC and 20% each of CV and VC sounds. These three types of speech

TABLE 2.1: Types of Syllables in Telephone Conversations in Percent after Table 13 of Fletcher (1953)

SYLLABLE TYPE	OCCURRENCE (%)
V	9.7
VC	20.3
CV	21.8
CVC	33.5
VCC	2.8
CCV	0.8
CVCC	7.8
CCVC	2.8
CCVCC	0.5
	100

sounds have been shown to compose 76% of all telephone speech (see Table 2.1:). This use of balanced *nonsense* speech sounds approximately maximizes the *entropy* of the corpus. This was an important method, first used in about 1921, to control for context effects, which were recognized as having a powerful influence on the recognition score P_c.

The speech corpus was held constant across experiments to guarantee that the source entropy was constant. Even though information theory had not yet been formally proposed, the concepts were clear to those doing the tests. The rules were set by Campbell, Fletcher, and Stewart, all of whom were trained mathematically and were highly sophisticated in such detailed matters.

The articulation test crew consisted of 10 members, with 1 member of the crew acting as a *caller*. A typical test is shown in Fig. 2.2. This record is from March 1928, and the testing condition was lowpass filtering at 1500 Hz. The

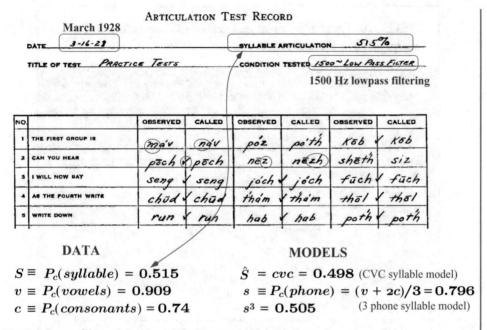

FIGURE 2.2: Typical test record for the 1928 Western Electric research Laboratory speech intelligibility testing method.

sounds were typically varied in level to change the signal-to-noise ratio to simulate the level variations of the network. Thus three types of distortions were simultaneously used: lowpass filtering, highpass filtering, and a variable SNR (Fletcher, 1995).

What they found: In the example shown in Fig. 2.2, the percent correct for syllables is 51.5%. The test consisted of the caller repeating context neutral (ZP) sentences, such as "The first group is *na'v*" and "Can you hear *pōch*." In the first presentation, the syllable was incorrectly heard as "ma'v," making an error on the initial consonant. In the second presentation the syllable is correctly heard. All the initial consonants, vowels, and final consonants, were scored and several measures were computed.

The syllable probability correct $[S \equiv P_c(\text{syllable})]$ was found to be 51.5% correct, as shown in the upper right corner of the score sheet. The vowel recognition score v was 90.9%. The average of the initial c_i and final c_f consonant score $c = (c_i + c_f)/2$ was found to be 74%. These numbers characterize the raw data. Next the data is modeled, as shown on the right lower portion of the figure. The *mean-CVC-syllable* score is modeled by the triple product

$$\widehat{S} = cvc. \tag{2.1}$$

From many measurements it was found that these models did a good job of characterizing the raw data (Fletcher, 1995, pp. 175, 178, 196–218). Based on many records similar to the one shown in Fig. 2.2, they found that the *average MaxEnt CVC phone recognition*, defined by

$$s \equiv (2c + v)/3, \tag{2.2}$$

did a good job of representing MaxEnt CVC syllable recognition, defined by

$$S_3 \equiv cvc \approx s^3. \tag{2.3}$$

Similarly, MaxEnt CV and VC phone recognitions were well represented by

$$S_2 \equiv (cv + vc)/2 \approx s^2. \tag{2.4}$$

These few simple models worked well over a large range of scores, for both filtering and noise (Rankovic, 2002). Note that these formulae only apply to MaxEnt speech sounds, *not* meaningful words.

The above models are necessary but not sufficient to prove that the phones may be modeled as being independent.[4] Namely such models follow if independence is assumed, but demonstrating their validity experimentally does not prove independence. To prove independence, all permutations of element *recognition* and *not-recognition* would need to be demonstrated. The closest we have to such an analysis is the work of Bronkhorst *et al.* (Bronkhorst *et al.*, 1993, 2002), as discussed in Chapter 3.

---

[4] Independence is defined as the condition that the joint probability is the product $P(A, B) = P(A)P(B)$ for all sets $\{A, B\}$ for which $P(A, B)$ is defined (Wozencraft and Jacobs, 1965, Ch. 2).

The exact specifications for the tests to be modeled with these probability equations are discussed in detail in (Fletcher, 1929, pp. 259–262). A most interesting issue is the vowel to consonant ratio (Fletcher, 1995, Eq. (15.8), p. 283)

$$\lambda(SNR) \equiv v/c. \tag{2.5}$$

Fletcher went to some trouble to discuss the effect of this ratio on the average phone score s (this key argument is rarely, if ever, acknowledged), and showed that λ has a surprisingly small effect on s. These observations might be important in applications of AI theory to various languages if λ were significantly different from that for English. Another implication is that this insensitivity may reflect the much higher rate of recognition of vowels over consonants at moderate SNRs. Unfortunately these questions about $\lambda(SNR)$ remain largely unexplored.

In summary, the Fletcher *et al.* results were an important first step. The key result was their finding that the average phone score Eq. (2.2) is an important statistical measure, useful when modeling syllable scores, as in Eq. (2.3) or Eq. (2.4).

2.1.1   Articulation Index Theory

Given the successful application of the average phone score Eq. (2.2), Fletcher immediately extended the analysis to account for the effects of filtering the speech into bands (Fletcher, 1921, 1929). This method later became known as *articulation index* theory, which many years later developed into the well known ANSI 3.2 AI standard. To describe this theory in full, we need more definitions, as provided by Table 2.2.

The basic idea was to vary the SNR *and* the bandwidth of the speech signal, in an attempt to idealize and simulate a telephone channel. Speech would be passed over this simulated channel, and the articulation $P_c(\alpha, f_c)$ measured. The parameter α is the gain applied to the speech, used to vary the SNR, given a fixed noise level. The SNR depends on the *spectral level* (the power in a 1 Hz bandwidth, as a function of frequency) of the noise and α. The consonant and vowel articulation [$c(\alpha)$ and

TABLE 2.2: Table of Definitions Required for the Articulation Index Experiments

SYMBOL	DEFINITION
α	Gain applied to the speech
$c(\alpha) \equiv P_c(\text{consonant}\vert\alpha)$	consonant articulation
$v(\alpha) \equiv P_c(\text{vowel}\vert\alpha)$	vowel articulation
$s(\alpha) = [2c(\alpha) + v(\alpha)]/3$	Average phone articulation for CVC's
$e(\alpha) = 1 - s(\alpha)$	Phone articulation error
f_c	Highpass and lowpass cutoff frequency
$s_L(\alpha, f_c)$	s for lowpass filtered speech
$s_H(\alpha, f_c)$	s for highpass filtered speech
$S(\alpha, f_c)$	Nonsense syllable (CVC) articulation
$W(\alpha, f_c)$	Word intelligibility
$\mathcal{I}(\alpha, f_c)$	Sentence intelligibility

$v(\alpha)$] and $s(\alpha)$ are functions of the speech level. The *mean phone articulation error* is $e(\alpha) = 1 - s(\alpha)$.

The speech was filtered by complementary lowpass and high-pass filters, having a cutoff frequency of f_c Hz. The articulation for the low band is denoted as $s_L(\alpha, f_c)$ and for the high band as $s_H(\alpha, f_c)$.

The syllable, word, and sentence intelligibility are $S(\alpha, f_c)$, $W(\alpha, f_c)$, and $I(\alpha, f_c)$, respectively.

Formulation of the AI: Once the functions $s(\alpha)$, $s_L(\alpha, f_c)$, and $s_H(\alpha, f_c)$ are known, it is possible to find relations between them. These relations, first derived by Fletcher in 1921, were first published by French and Steinberg (1947).

Fletcher's key insight here was to find a linearizing transformation of the results. Given the wideband articulation $s(\alpha)$, and the banded articulations $s_L(\alpha, f_c)$

and $s_H(\alpha, f_c)$, he sought a nonlinear transformation of probability AI, now called the *articulation index*, which would render the articulations additive, namely

$$AI(s) = AI(s_L) + AI(s_H). \tag{2.6}$$

While there was no guarantee that such a transformation might exist, his intuition was correct. This formulation payed off handsomely.

The function $AI(s)$ was determined empirically. It was found that the data for the MaxEnt sounds closely follows the relationship

$$\log(1 - s) = \log(1 - s_L) + \log(1 - s_H), \tag{2.7}$$

or in terms of error probabilities

$$e = e_L e_H, \tag{2.8}$$

where $e = 1 - s$, $e_L = 1 - e_L$, and $e_H = 1 - s_H$. These findings suggest $AI(s)$ is of the form

$$AI(s) = \frac{\log(1 - s)}{\log(e_{min})}. \tag{2.9}$$

This normalization parameter e_{min} is the minimum error, while s_{max} is the maximum value of s, given ideal conditions (i.e., no noise and full speech bandwidth), with $e_{min} \equiv 1 - s_{max}$. Solving Eq. (2.9) for s gives

$$s(AI) = 1 - e_{min}{}^{AI}. \tag{2.10}$$

This equation is the fundamental relationship that specifies the average phone score for MaxEnt C,V sounds, in terms of the AI. Again it must be emphasized that Eq. (2.10) only holds for MaxEnt speech sounds.

The total error $e = e_{min}{}^{AI}$ in Eq. (2.10) was represented by Fletcher as $e = 10^{-AI/0.55}$. Both expressions are exponential in AI, differing only in the choice of the base $\left(e_{min} = 10^{(-1/0.55)}\right)$.

For much of the Bell Labs work $s_{max} = 0.9848$ (i.e., 98.5% was the maximum articulation), corresponding to $e_{min} = 0.015$ (i.e., 1.5% was the minimum articulation error) (Rankovic and Allen, 2000; MM-3373, Sept. 14, 1931, J.C. Steinberg; Fletcher, 1995, p. 281) and Galt's notebooks (Rankovic and Allen, 2000).

Fletcher's simple two-band example (Fletcher, 1995, p. 281) illustrates Eq. (2.8): If we have 100 spoken sounds, and 10 errors are made while listening to the low band and 20 errors are made while listening to the high band then

$$e = 0.1 \times 0.2 = 0.02, \tag{2.11}$$

namely two errors will be made when listening to the full band. Thus the wideband articulation is 98% since $s = 1 - 0.02 = 0.98$. The wideband MaxEnt CVC syllable error would be $S = s^3 = 0.941$.

In 1921 Fletcher, based on results of J.Q. Stewart, generalized this two-band (Eq. 2.8) case to $K = 20$ bands:

$$e = e_1 e_2 \cdots e_k \cdots e_K, \tag{2.12}$$

where $e = 1 - s$ is the wideband average error and $e_k \equiv 1 - s_k$ is the average error in one of K bands. Formula 2.12 is the basis of the *articulation index*.

The K band case has never been formally tested, but was verified by working out many examples, as discussed in Section 2.7. The number of bands $K = 20$ was an empirical choice that was determined after many years of experimental testing. Each of the bands was chosen to have an equal contribution to the articulation. The average of the specific articulation index AI_k over the 20 bands gives the total articulation index AI, namely

$$AI = \overline{AI_k} \equiv \frac{1}{K} \sum_{k=1}^{K} AI_k, \tag{2.13}$$

which is the generalization of Eq. 2.6 to K bands.

The number $K = 20$ was a compromise that probably depended on the computation cost as much as anything. Since there were no computers, too many bands was prohibitive with respect to computation. Fewer bands were insufficient.

Eventually Galt discovered that articulation bands, defined as having equal articulation, were proportional to cochlear critical bands (French and Steinberg,

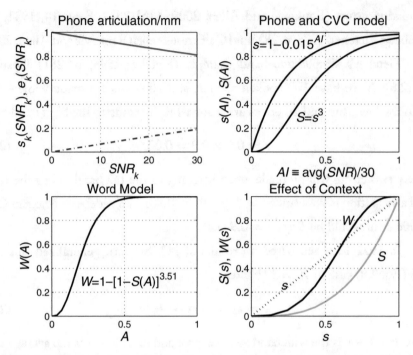

FIGURE 2.3: Typical results for the French and Steinberg AI model, as defined by Eqs. (2.12)–(2.16) and in Fig. 3.1. With permission from Allen (1994).

1947, p. 92). Each of these K articulation bands corresponds to approximately 1 mm along the basilar membrane (Fletcher, 1940, 1953, p. 172, Allen, 1996). Assuming a cochlear critical band corresponds to about 40 hair cells (Allen, 1996), and each hair cell is about 12 μm, one articulation band is about 2 critical bands wide.[5] When the articulation is normalized by the critical ratio, as a function of the cochlear tonotopic axis, it was found that the articulation density is constant (per critical band). This relation has been further quantified in Allen, 1996 in terms of a plot of the ratio of the articulation bandwidth over the critical ratio, as shown in Fig. 2.4.

---

[5] As a cross check, Shera *et al.* (2002) estimated the human critical band to be $\Delta_f \approx 143$ Hz at 2 kHz namely, $Q_{ERB}(2kHz) \equiv 2000/\Delta_f \approx 14$. This bandwidth corresponds to 0.11 octaves, given by $\log_2[(2000 + 0.5\Delta_f)/(2000 - 0.5\Delta_f)]$. The spatial extent corresponding to 0.11 may be determined by multiplying by the slope of the human cochlear map, which is 5 mm/oct (Greenwood, 1990). Thus the critical spread Δ_x at 2 kHz is 0.5 mm.

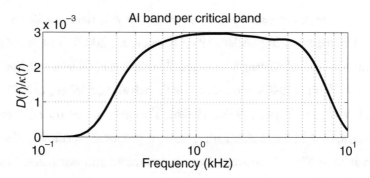

FIGURE 2.4: This figure shows the ratio of the articulation index bandwidth for constant articulation to critical ratio. It was generated from the data given in Fletcher (1995, Fig. 177, p. 289) for $D(f)$ divided by the critical ratio $\kappa(f)$ as given in Fig. 121, p. 167. Since the ratio is approximately constant over the frequency range from 300 Hz to 7.5 kHz, it shows that the articulation bandwidth is proportional to the critical bandwidth. With permission from Allen (1996).

2.2   FRENCH AND STEINBERG (1947)

In 1947 French and Steinberg provided a valuable extension of the formula for the band errors by relating e_k (the kth band probability of error) to the band SNR, SNR_k, by the relation[6]

$$e_k = e_{\min}{}^{AI_k(SNR_k)/K},$$ (2.14)

with $e_{\min}$ being the minimum wideband error, under ideal conditions.

In each cochlear critical (i.e., articulation) band, the signal and noise is measured, and the long term ratio is computed as

$$snr_k \equiv \frac{1}{\sigma_N(\omega_k)} \left[\frac{1}{T} \sum_{t=1}^{T} \sigma_s^2(\omega_k, t) \right]^{1/2},$$ (2.15)

where $\sigma_s(\omega_k, t)$ is the short-term RMS of a speech frame and $\sigma_N(\omega_k)$ is the noise RMS, for frequency band k. The time duration of this frame impacts the definition of the SNR, and this parameter must be chosen to be consistent with a cochlear analysis of the speech signal. It seems that the best way to established this critical

---

[6] This relation is related to Eq. (10a) of the 1947 French and Steinberg paper, but was first stated in this form by Allen (1994).

duration is to use a cochlear filter bank, which to this date is still an uncertain quantity of human hearing (Allen, 1996; Shera *et al.*, 2002). The standard engineering method for calculating a perceptually relevant SNR was specified in 1940 by Dunn and White (1940), and refined by French and Steinberg (1947), Fletcher and Galt (1950), and Kryter (1962a, 1962b). This may still be the best way, until a new standard, based on a cochlear filter bank, is proposed, verified, and accepted.

Each band SNR_k is converted to dB, is limited and normalized to a range of −6 to 24 dB, and normalized by 30, thereby defining the *specific articulation*, denoted $AI_k(SNR_k)$:

$$AI_k = \begin{cases} 0 & 20\log_{10}(snr_k) < -6 \\ 20\log_{10}(snr_k)/30 & -6 < 20\log_{10}(snr_k) < 24 \\ 1 & 24 < 20\log_{10}(snr_k). \end{cases} \qquad (2.16)$$

The threshold of speech perception is close to an SNR_k of −6 dB ($snr_k = 0.5$) (French and Steinberg, 1947). The factor 30 in Eq. (2.16) results from the audibility of speech over a 30 dB dynamic range in a given articulation band (French and Steinberg, 1947, Fig. 4, p. 95). The right-hand side of this formula is a restatement of the straight line approximation of Fig. 4 of French and Steinberg (p. 95). The left-hand side is defined in their Fig. 21.

The basic idea of this formula is that when the SNR_k is less than threshold (i.e., −6 dB) within each cochlear critical band, the speech is undetectable. When SNR_k is greater than 24 dB, the noise has no effect on the intelligibility. Between −6 and +24 dB the AI_k is proportional to $\log(SNR_k)$. This formula ignores the upward spread of masking, and is not valid when this important effect is triggered, for example when the speech is low pass filtered and amplified.

Merging the formula for the band errors Eq. (2.12) with that for the specific AI Eq. (2.16), the total error may be related to the average specific AI, Eq. (2.13), via Eq. (2.14), leading to

$$e = e_1 e_2 \cdots e_K = e_{\min}{}^{AI_1/K} e_{\min}{}^{AI_2/K} \cdots e_{\min}{}^{AI_K/K} = e_{\min}{}^{AI}. \qquad (2.17)$$

Since $s = 1 - e$, Eq. (2.10) follows, as required. Note that as $SNR_k \rightarrow 30$ dB in every band, $AI \rightarrow 1$ and $s \rightarrow s_{max}$. When $SNR_k \rightarrow -6$ dB in all the bands, $AI \rightarrow 0$ and $s \rightarrow 0$.[7] This formula for $s(AI)$ has been verified many times, for a wide variety of conditions (see Section 2.7)

2.3 EFFECTS OF CHANCE AND CONTEXT

There are two major problems that the early Bell Labs studies did not address – in fact they designed around these issues. The first is accounting for *chance* performance (guessing), and the second is the role of *context*. A complete theory of HSR must accurately account for their influence.

When dealing with chance and context one needs an understanding of the basic aspects of Claude Shannon's theory of information (Shannon, 1948). Information theory deals with chance with the concept of *entropy*, and with context effects via *conditioned probability*. To appreciate and understand these tools intuitively we need a brief introduction to some practical issues in modeling articulation and intelligibility with probability theory. In the next section three key topics are discussed: entropy, channel capacity, and probability composition laws.

Chance performance plays its largest role at very low *SNR*s, where all of the signal channels have error 1. Chance may be modeled as a side channel that reduces the error to the maximum entropy condition. Context is modeled using conditional probability, and is most important at high *SNR*s. When the rules of chance dominate, context can play no role.

One unified way of seeing the relations of probability is through the multi-dimensional diagram shown in Fig. 2.5. This figure gives us a visual and uniform way of dealing with the probability relations that describe speech sound recognition. In this figure each *channel* is represented as a dimension. Because all possible outcomes of any trial may be represented by a point in the space, the total volume must be 1. In the example shown in the figure, binary outcomes are represented,

---

[7] The symbol $\rightarrow$ is read "goes to."

CONSERVATION OF PROBABILITY

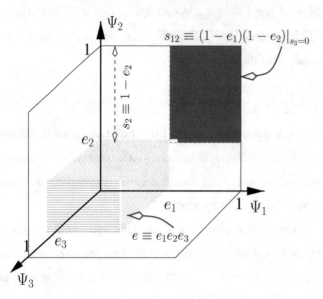

FIGURE 2.5: This figure provides the conservation of probability interpretation of articulation error probability. It may be thought of as a 3^d diagram, and it can be helpful, when thinking about the AI model.

true or false, which represent the probability of an event being recognized, or not. As an example, think of an experiment where three biased coins are tossed, and we wish to represent the outcome of every trial (tossing each of the three coins, with outcomes restricted to the eight corners of the cube

$$\{\delta_i, \delta_k, \delta_k\} \equiv \frac{\{1 \pm 1, 1 \pm 1, 1 \pm 1\}}{2}. \tag{2.18}$$

In this example, s_1 might represent the probability of seeing heads (1) on the first coin when a head was tossed, while e_2 would then be the probability of seeing heads on the second coin when a tail (0) was tossed.

As a more relevant example, the probability of not recognizing the speech events in any of the cochlear filter channels is e, as given by Eq. (2.12), which represents a 20-dimensional volume (which cannot be drawn, but we can imagine).

A projection defines a conditional probability, such as $P(x_1, x_2 \mid x_3 = 0)$, as shown by s_{12}.

Most of what we have seen, and what we will need, for describing the probability relations in HSR, may be related back to Fig. 2.5.

2.3.1   Composition Laws

There is an important theme that threads all of the probability relations Eqs. (2.1)–(2.12). In some cases the statistical models are formed from products of *articulations*, such as Eq. (2.3) and Eq. (2.4), while in other cases they are formed from products of *articulation errors*, such as in Eq. (2.8) and Eq. (2.12). What determines which to use, the products of probability correct, as in Eq. (2.3), or probability of error, as in Eq. (2.12)?[8]

In Fig. 2.5 we see an interpretation of these various products of probability in terms of a volume in probability space. If the entire space is normalized to have a volume of 1, then each of the various types of products defines a subvolume. The sum of all of these subvolumes adds to 1, leading to the view of probability as a conservation law.

One way to describe these rules intuitively is in terms of *sequential* versus *parallel* processing. When the processing is *sequential*, one must multiply articulations, and when the processing is *parallel*, one must multiply articulation errors.

Sequential processing: When one needs to recognize *every* member of a string having no context we shall call this *sequential processing*: all the elements must be correct for the result to be correct. Sequential processing always drives the score *down* (a single error makes the entire result wrong).

An example of *sequential processing* is the case where all the phones must be recognized to successfully recognize the composite MaxEnt word. Context cannot

---

[8] R. H. Galt raises this question in his notebooks (Galt, 1946) where he points to Fry's interesting book on Probability Theory and *The first and second law of composition of probability* (Fry, 1928).

help in this case since the phones are independently drawn, and all sequential *partial events* must be recognized to recognize the *compound event*. This requires that P_c be given by a product of articulations rule, as in Eq. (2.3).

If all the phones are at the chance level, then the syllable must be at chance. Thus probability correct for sequential processing can never go to zero, because that would require a P_c that is below chance.

Parallel processing: When one only needs to recognize a single member of many elements, we shall call this *parallel processing*. Thus the listener has many (i.e., parallel) opportunities to get the correct answer. A primary example is Eq. (2.12) corresponding to the parallel processing via K different independent channels. As a second example, if there is one unique sound (i.e., a vowel) that distinguishes two words, or a salient word in a sentence that is identified, then the result is scored correct, and parallel processing applies. A third example might be

$$P_c \equiv 1 - (1 - p_c)^k, \tag{2.19}$$

where p_c is the probability correct of k identical independent parallel channels.

One problem with Eq. (2.19) is that it does not account for chance guessing. A small modification, by adding a chance channel, can fix this. If 1 out of N possibilities are being considered (as in close set testing), and the sounds are uniformly chosen, then

$$P_c = 1 - \left(1 - \frac{1}{N}\right)(1 - p_c)^k. \tag{2.20}$$

This model approaches chance level ($P_c \to 1/N$) as $p_c \to 0$, rather than zero.

For parallel processing the error may be zero, corresponding to the case where the signal is clearly detectable in a single channel. However it is more likely that the coincidence of many channels define each event, implying a nonzero error for any single channel. When many channel error probabilities are multiplied together, the total error probability may become vanishingly small. For example, if each of

10 independent contiguous channels have an error probability of 0.5 (50%), then the probability of error of the group estimate will be $0.5^{10} \approx 10^{-3}$ (0.1%).

An important (and historically the first) example of *parallel processing* is Eq. (2.8). In Eq. (2.12) many (even most) of the channels may have no information, and thus have an error approaching 1. If one channel gives an error free signal (e.g., if $e_3 = 0$), then that single channel dominates the final result.

While it is trivially obvious, it is worth emphasizing, that when one takes a number that is less than one, to a power that is greater than one, the result becomes smaller. This means that *sequential processing* [i.e., Eq. (2.4) or Eq. (2.3)] reduces the score ($P_c^k < P_c$, $k > 1$), while *parallel processing* [i.e., Eq. (2.8) or Eq. (2.12)] increases the score ($P_e^j < P_e$, $j > 1$).

Boothroyd (Boothroyd, 1978; Boothroyd and Nittrouer, 1988) addresses the sequential vs parallel processing question in terms of two rules. When the articulations are multiplied, Boothroyd calls it "elements of wholes → wholes," which requires what he calls a "*j*-factor," as in

$$P_c \equiv p_c^j. \tag{2.21}$$

When articulation errors are multiplied, he views the situation as describing a mapping from "no context → context," which requires what he calls a "*k*-factor" as in Eq. (2.19). We shall describe Boothroyd's approach in detail in Section 3.1.

2.4   MILLER *ET AL.* CIRCA 1947–2001

In this section we shall deal with the important issues of entropy and chance, plus some restricted issues regarding context. George A. Miller was the first to explore the use of information theory in both HSR and human language processing (HLP). Miller and his colleagues raised and clarified these issues in some key speech papers. In one classic study Miller was the first to use *closed-sets* to control the entropy of the listening task. By doing this, it was possible to study the importance of chance as an independent variable. In a second classic study, he quantified the error patterns of the HSR *channel*, by measuring the confusion matrix for consonants. Following an in-depth review of this work, the data will be modeled using the tools of the articulation index, using parallel and sequential models.

The articulation index theory was developed by the telephone company, for network characterization. It was largely unknown outside AT&T until World War II, when voice communication became a matter of life and death.[9] Bell Labs was asked to participate in solving these communications problems, so Fletcher and his team went to Harvard to provide support. This meeting, of 31 people at Harvard on June 19, 1942, is documented in Galt's sixteenth notebook, starting on page 158 (Rankovic and Allen, 2000), and in the personal notes of Stevens about this meeting (Rankovic, personal communication).

The war effort involved teams at both Bell Labs and the Harvard Cruft Acoustics Lab, and one of the key players was a young man by the name of George Miller. Miller's contribution to our understanding of speech intelligibility was second to none. Following the war he wrote an interesting review paper summarizing the AI method (Miller, 1947a). Next he attempted to integrate Shannon's theory of information with Fletcher *et al.*'s articulation studies. These studies resulted in a book *Language and communication* (Miller, 1951), which reviews what was known at that time about speech intelligibility. In the same year, Miller *et al.* published a classic paper on the importance of the size of the set of sounds (i.e., the set entropy) to syllable classification (Miller *et al.*, 1951). He had (and still has) a good intuition for rooting out the problems, and drawing attention to them. His work during the war, and later with Heise and Lichten, are wonderful examples of this.

Closed- verses open-set testing: Although George Campbell (who also worked for AT&T) was the first to use the *closed-set test* (Campbell, 1910), all the subsequent Bell Labs work consisted of *open-set* tests, where chance is negligible (i.e., less than 2^{-11}). MaxEnt syllables were used to avoid word-context bias. With such tests

[9] The first application outside of the Bell System was pilot-navigator communications (Miller *et al.*, 1946; Miller and Mitchell, 1947).

every phone must be identified, and therefore sequential processing models apply. This is a maximum entropy test, implying a greater testing efficiency.

If only meaningful words are used in the open-set test, one unique key element (e.g., a salient sound) would be enough to narrow the choices, meaning *parallel processing* would then play a role, making the analysis difficult, or even impossible.

Open-set tests are quite difficult however, and are not particularly useful in the clinic, since they require highly trained person.

The essence of the *closed-set* test is that the subject has the full knowledge of the choices, *before* they are asked to make their selection. By explicitly show-ing the subject the possible choices, meaningful and MaxEnt-words are rendered equivalent. Thus the closed-sets task significantly reduces the role of word context. Assuming equally distributed choices, chance is given by the inverse of the number of choices; thus it is known, and is typically significant.

The down side of the closed-set task comes when meaningful words are exclusively used since the degrees of freedom (the number of relevant sounds per word) is unknown. Thus parallel processing must play a role, and modeling the results is more complex. As the *SNR* is increased, the number of distinct elements may be unknown. This means that the scoring will be difficult. However because the set size is small, it may not be impossible, depending on the nature of the corpus. Examples are useful to make this point.

Suppose your ask the subject to identify one of two words [*cat, dog*], in a closed-set task (the two words are known in advance). A coin is flipped to pick the word, which is then bandpass filtered, and masked with noise, such that it is hardly recognizable. Chance is 0.5. With such a closed set task (one of two words), you will only need to distinguish the vowel /a/ versus /o/. Alternatively if you can hear the high frequency release of the /t/ of cat, you will get the entire word right. Thus there are multiple strategies for guessing the word. This leads to modeling complications.

If the word *cat* were presented in an open-set task, you must identify each of the three sounds. If the word /cat/ were replaced with the meaningless CVC /dat/, the scores would likely not change, since all the audible cues would approximately remain the same.

2.4.1 Miller *et al.* (1951)

This study was the first to quantify task entropy using close-set testing. Starting from a master list of $N \approx 2048$ CVC words,[10] seven types of lists, of 2, 4, 8, 16, 32, 256, and 1000 words were prepared in advance. Denoted these lists $\text{List}_{n,l}$, containing 2^n words each, with $n = 1, 2, \ldots, 10$. A block of trials was a run from one of these List $_{n,l}$, where $l = 1, 2, 3, \ldots, L$. For example, when $n = 4$, the lists were $2^4 = 16$ words long. There are $2048 \times 2047 \times 2046 \times 2045$ such possible lists of 16 words. For the case of $n = 11$, one list includes *all* the words.

A word was chosen at random from a randomly chosen list, and 7 kHz bandwidth noise was added to the speech electrical waveform. The 1951 Miller *et al.* study used a wideband measure of the speech based on volts peak across the earphone. The bandwidth of the speech plus noise was determined by the earphones, to be about 4.5 kHz. The values of SNR chosen were between -21 and $+18$ dB in 3 dB steps. The subjects were shown the chosen list, and after listening to the processed sample, they identified one word on the list that most closely matched what they heard. After a block of trials at each value of n, the average score was computed for that block. This process was continued until sufficient data at each value of n and SNR had been collected.

The Miller *et al.* experimental results, $P_c(SNR, \mathcal{H})$, are summarized in Fig. 2.6. The data are re-plotted in a different format in Fig. 2.7, where the entropy of the task $\mathcal{H}$ is the abscissa and the ordinate is percent correct ($100 \times P_c$), plotted on a log scale, with SNR as the parameter. Chance is bounded by the line labeled $2^{-\mathcal{H}}$ (since the words were independently chosen from the list, without bias) corresponding to an SNR of about -21 dB.

---

[10] Some typical words are bat, kiss, all, void, root. The list was actually 1000 words, but it's effective size was about 2000, Personal communication, G. Miller (2000).

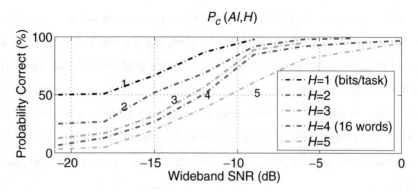

FIGURE 2.6: The $PI(SNR,\mathcal{H})$ functions are shown for an $\mathcal{H}$ of 1 to 5 bits. The 1-bit task (two words) chance is 1/2, which is reached at an SNR of -17 dB. For the 16 word 4-bit task, chance is reached at -21 dB SNR.

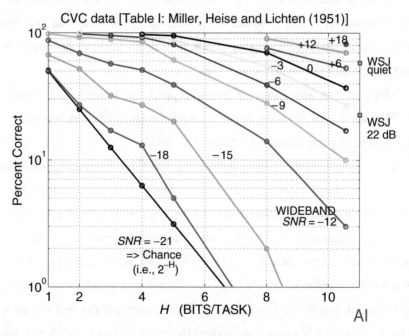

FIGURE 2.7: In 1951 Miller *et al.* measured the word intelligibility score W of a word spoken randomly from a word list, as a function of the list's size N and the SNR. This chart displays these results on a log-log scale, along with chance performance for the task. The chart relates the entropy $H \equiv \log_2(N)$ (in bits/task) and the SNR for $W(H, SNR)$, with the SNR as a parameter. Two ASR scores for the *Wall Street Journal* (WSJ) task of (Lippmann, 1996) have been added for reference.

Below about -21 dB SNR, the subjects heard only noise. At an *SNR* of -18 dB, the subjects perform slightly above chance for the 3–5 (i.e., 8–32 word lists) bit tasks, and the performance is maximum (relative to chance) for $\mathcal{H} = 4$ (16 words). At this *SNR* it is likely that only the CVC's centered vowel can be heard. There are about 20 centered vowels.[11] Since the number of words (16) is less than the number of vowels (20), the number of repeated vowels, given two words, will be small, resulting in a greatly increased chance, given the vowel, for guessing the word.

For comparison, two points from an ASR experiment on the Wall Street Journal task have been added to the figure (Lippmann, 1996). These ASR results are much worse than the Miller *et al.* HSR measures at the same conditions. Note the hypersensitivity to noise, where a slight change in SNR results in a dramatic drop in the score, from 60% in quiet to 21% at $+22$ dB SNR. A noise of this small magnitude would have no affect on HSR.

Many of the key issues of human speech recognition are captured by MHL51's experiment. This figure summarizes the trade-off between entropy and SNR and chance performance in a rather nice way.

The use of closed-set testing, first explored by this experiment, fundamentally changed the science of speech articulation testing. First, the entropy of the task is controlled. Second, with closed–set testing, chance of correct identification is known (e.g., $P_c(\text{chance}) = 2^{-H}$). Knowing the chance response should allow one to more precisely model the results. However, before we can make such models we must know the distribution of the number of common Cs and Vs on the list, since this distribution will determine $P_c(SNR)$. For example, for the case of two words, if all the Cs and Vs differ (e.g., cat vs. dog) the result be be very different from the case were only the vowel is varied (e.g., cat vs. cut).

In the open-set Bell studies, the subjects were necessarily highly trained, and they needed to know phonetic symbols. Much less practice is required for the

---

[11] Based on an early Bell Labs survey, as summarized by Fletcher, the vowels have an entropy of about 4.3 bits (Allen, 1994).

subjects in the MHL51 closed set task, for the same accuracy, when meaningful words are used. In my view, when done properly, this type of testing should provide results that are as accurate as open-set tests using MaxEnt-words, but much easier to administrate, and much broader in their ability to evaluate speech sound perception. Coupled with the use of computers, such tests could become quite useful today.

Modeling closed set testing (MHL51):   As the entropy of the list increases from 1 bit to 11, the effective number of phones that cue the word must change from 1 to 3 (assuming CVC words). This must show up in a change of the slope of the log-error curve of $P_c(SNR, \mathcal{H})$, after correcting for chance. This changing slope effect may be seen in Fig. 2.8 where

$$P_e(SNR, \mathcal{H}) = \frac{1 - P_c(SNR, \mathcal{H})}{1 - 2^{-\mathcal{H}}} \qquad (2.22)$$

is shown, with the entropy $\mathcal{H}$ as the parameter.

At higher SNR values, most of the consonants are heard. The end result will be a large performance increase in the closed-set PI functions as $\mathcal{H}$ decreases, as there will be many cues available, and a decreasing number of sounds to chose from. For $\mathcal{H} \leq 5$, all the curves above 0 dB SNR saturate, showing a "ceiling effect." Another way of saying this is that for 32 or less words, with an SNR of greater than −6 dB, human performance is close to perfect. In this region, performance will be limited by secondary factors such as production errors, attention, motivation, and memory. Because of the orthographic side information, only one unique combination of sounds must be identified to get the entire word right.

The dashed curve in Fig. 2.8 is the cube of the $\mathcal{H} = 5$ log-error curve. It almost matches the $\mathcal{H} = 1$ curve. This shows that, at least approximately, the degrees of freedom in the 32 word case are about three times greater than in the 2 word case. This is is what one would expect to happen, for CVC sounds having 3 degrees of freedom, as shown by Eq. (2.3). This parallel processing (k-factor)

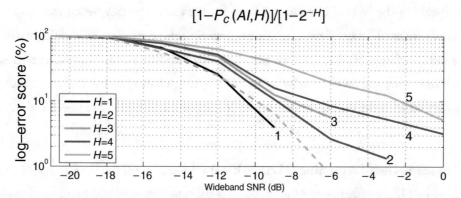

FIGURE 2.8: The same data shown in Fig. 2.6 is replotted as a log error, corrected for chance, namely $[1 - P_c(SNR, \mathcal{H})]/(1 - 2^{-\mathcal{H}})$ is plotted, as a function of SNR, on a log scale. The dashed curve is the cube of the $\mathcal{H} = 5$ curve, which comes close the $\mathcal{H} = 1$ curve, as discussed in the text.

model nicely characterizes the $\mathcal{H} = 1 \ldots 5$ data, since

$$\left(\frac{1 - P_c(SNR, 1)}{1 - 2^{-1}}\right) \approx \left(\frac{1 - P_c(SNR, 5)}{1 - 2^{-5}}\right)^3 \tag{2.23}$$

A parallel-processing model is further explored in Fig. 2.9. By using log-log coordinates the k factor shows up as a change in slope. If the lines are straight, as they are up to $\mathcal{H} = 4$, the k factor explains the data. The model works approximately for $\mathcal{H} \leq 7$. This model does not explain the $\mathcal{H} = 8$ and $\mathcal{H} = 10.5$ bit data, but for those data the lists were too long for the subjects to use them, and therefore word context (i.e., intelligibility) likely played an important role in those cases.

A second context result is for the case of the digits, as shown in Fig. 2.10. This data shows that the parallel model works for the digits using a power of 1.5, as discussed in the figure caption.

One thing these collective results do not tell us is *how* humans process the speech signal events to extract events. The first quantitative insight into that problem was provided by Miller and Nicely (1955).

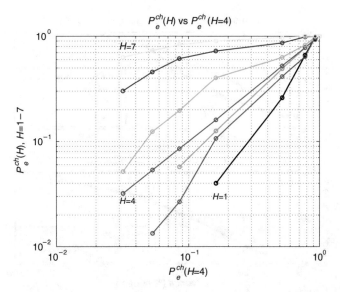

FIGURE 2.9: The same data shown in Fig. 2.6 is re-plotted as a log error, corrected for chance, namely if $P_e^{ch}(SNR, \mathcal{H}) \equiv [1 - P_c(SNR, \mathcal{H})]/(1 - 2^{-\mathcal{H}})$, then this plot is $P_e^{ch}(\mathcal{H})$ as a function of $P_e^{ch}(\mathcal{H} = 4)$. To the extent that the lines are straight (they are not for $\mathcal{H} = 5, 7$), one curve is a power law of the other. Namely a parallel processing ("k-factor") model does a reasonable job of accounting for the $\mathcal{H} = 1 - 4$, but less so for $\mathcal{H} = 5 - 7$. For the higher entropy cases, something else is going on.

2.4.2 Miller and Nicely (1955)

A few years later George Miller and Patricia Nicely published one of the most insightful and influential HSR papers to date. In many ways this is an extremely limited study. The five participants were American born women, who likely had a uniform north American dialect. Four served as listeners, with a randomly chosen one of the five serving as the live talker (recordings or phonograph records of the speech were not used, as was the case for the Bell studies). The database, was entirely CV speech, with a single vowel, always the /a/ as in the word father.

The results of this study clearly demonstrate what one might light-heartedly call "the quantum mechanics" of speech perception, i.e., that phone recognition is grounded on *hierarchical categorical discriminations*. This conclusion follows from an analysis of the Miller Nicely confusion data.

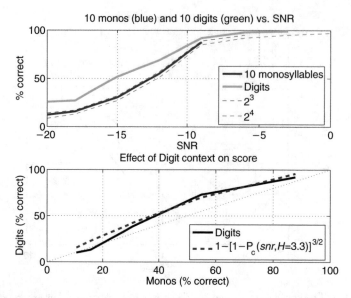

FIGURE 2.10: In the upper panel the 10-digit PI function is compared to the 8 and 16 word case, shown by the dashed curves. The solid curve, between the dashed curves, shows the PI(*SNR*) function for the 10 monosyllable case, determined by a linear interpolation between the 8 and 16 word cases. The second solid curve, above the dashed curves, represents the 10 digits. In the lower panel the digit PI function is compared to the 10-digit monosyllables. This shows a slight lifting of the digits over the 10 word monosyllables. The digits are then modeled with a parallel processing model having a *k*-factor of $k = 3/2$ [Eq. (2.19)], as shown by the dashed line. The model might fit better at low snr values if the mono's $P_c(SNR, \mathcal{H})$ values were first corrected for chance, and then reapply the chance model after taking the error to the 1.5 power. This correction was not verified.

Representations of the confusion matrix (CM): Fig. 2.11 shows a typical Miller–Nicely (MN55) consonant-vowel (CV) *confusion matrix* or *count matrix* CM for wideband speech (0.2–6.5 kHz), at a SNR of −6 dB (Miller and Nicely, 1955, Table III). The 16 consonants were presented along with the vowel /a/ as in father (i.e., the first three sounds were [/pa/, /ta/, /ka/]). After hearing one of the 16 CV sounds as labeled by the first column, the consonant that was reported is given as labeled along the top row. This array of numbers form the basic CM, denoted $C_{s,h}$, where integer indices s and h (i.e., "spoken" and "heard") each run between 1 and 16. For example, /pa/ was spoken 230 times (the sum of the counts in the first row),

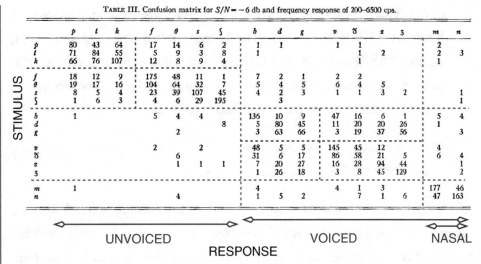

TABLE III. Confusion matrix for S/N = −6 db and frequency response of 200–6500 cps.

STIMULUS

	p	t	k	f	θ	s	ʃ	b	d	g	v	ð	z	ʒ	m	n
p	80	43	64	17	14	6	2	1	1		1	1			2	
t	71	84	55	5	9	3	8	1			1	2			2	3
k	66	76	107	12	8	9	4			1		1			1	
f	18	12	9	175	48	11	1	7	2	1	2	2				
θ	19	17	16	104	64	32	7	5	4	5	6	4	5		1	
s	8	5	4	23	39	107	45	4	2	3	1	1	3	2		1
ʃ	1	6	3	4	6	29	195			3		1	1		1	1
b	1			5	4	4		136	10	9	47	16	6	1	5	4
d							8	5	80	45	11	20	20	26	1	
g						2		3	63	66	3	19	37	56		3
v			2				2	48	5	5	145	45	12		4	
ð				6				31	6	17	86	58	21	5	6	4
z					1	1	1	7	20	27	16	28	94	44		1
ʒ								1			3	8	45	129		2
m	1							4			4	1	3		177	46
n				4				1	5	2	7	1	6		47	163

UNVOICED ↔   VOICED   NASAL

RESPONSE

FIGURE 2.11: Typical Miller–Nicely confusion (or count) matrix (CM) $\mathcal{C}$, from Table III at −6 dB SNR. Each entry in the matrix $\mathcal{C}_{s,h}$ is the subject response count. The rows correspond to the *spoken* CVs, each row representing a different consonant, from $s = 1, \ldots, 16$. The columns correspond to the *heard* CVs, each column representing a different consonant, from $h = 1, \ldots, 16$. The common vowel /a/, as in father, was used throughout. When the 16 consonants are ordered as shown, the count matrix shows a "block-symmetric" partitioning in the consonant confusions. In this matrix there are three main blocks delineated by the dashed lines, corresponding to UNVOICED, VOICED, and NASAL. Within the VOICED and UNVOICED subgroups, there are two additional symmetric blocks, corresponding to AFFRICATION and DURATION, also delineated with dashed lines.

and was reported heard 80 times ($\mathcal{C}_{1,1}$), while /ta/ was reported 43 times ($\mathcal{C}_{1,2}$). For Table III the mean row count was 250, with a standard deviation of 21 counts.

When the sounds are ordered as shown in Fig. 2.11, they form groups, identified in terms of hierarchical clusters of *articulatory features*. For example, the first group of sounds 1–7 correspond to UNVOICED, group 8–14 are VOICED, and [15,16] are NASAL and VOICED.

At an SNR of −6 dB, the intraconfusions (within a group) are much greater than the inter-confusions (between groups). For example, members of the group 1–7 (the UNVOICED sounds) are much more likely to be confused among themselves, than between the NONNASAL–VOICED sounds (8–14), or the NASAL sounds (15,16). The NASAL are confused with each other, but rarely with any of the other sounds 1–14. Two of the groups [sounds (1–7) and (8–14)] form subclusters.

A major conclusions of the Miller–Nicely paper is

This breakdown of the confusion matrix into five smaller matrices ... is equivalent to ... five communication channels

The term *communication channels* is being used in the Shannon sense. A *Shannon channel* is a "digital pipe" that introduces classification errors (i.e., confusions) due to noise. The *1-bit asymmetric channel* may be characterized by a matrix of the form

$$
\begin{array}{c|cc}
 & 0 & 1 \\
\hline
0 & p & 1-p \\
1 & 1-q & q
\end{array}
\tag{2.24}
$$

where $1 - p$ and $1 - q$ are the probabilities of the binary transmission errors, corresponding to a 0 or a 1. It is $p(SNR)$ that specifies the relation between the sound classifications errors and the analog SNR. The articulation for this example matrix would be $P_c(SNR, 1) = (p + q)/2$. The *1-bit symmetric channel* corresponds to the case of $q = p$.

Based on the block errors in the CM (Fig. 2.11), Miller and Nicely suggested that it was the misperception of basic speech features that account for the errors. They offered the *feature* classification scheme defined in Fig. 2.12. In this chart each consonant has a unique binary [$\log_2(3)$ bits in the case of place] representation.

To the extent that the CM is block-symmetric, the events identified in Fig. 2.11 seem to represent independent communication channels. More precisely, Miller and Nicely say.

the impressive thing to us was that ... the [binary] features were perceived almost independently of one another.

TABLE XIX. Classification of consonants used to analyze confusions.

Consonant	Voicing	Nasality	Affrication	Duration	Place
p	0	0	0	0	0
t	0	0	0	0	1
k	0	0	0	0	2
f	0	0	1	0	0
θ	0	0	1	0	1
s	0	0	1	1	1
ʃ	0	0	1	1	2
b	1	0	0	0	0
d	1	0	0	0	1
g	1	0	0	0	2
v	1	0	1	0	0
ð	1	0	1	0	1
z	1	0	1	1	1
ʒ	1	0	1	1	2
m	1	1	0	0	0
n	1	1	0	0	1

FIGURE 2.12: This figure shows the 5-event classification scheme of Miller–Nicely, Table XIX. Each of the sounds was assigned a binary event (in the case of "Place," the scheme required more than 1 bit). Today the term feature is widely used, and means many things. For this reason the term *event* is preferred when referring to Miller–Nicely's "features."

The MN55 data has been the inspiration for a large number of studies. The sound grouping has been studied using multidimensional scaling, which has generally failed in providing a robust method for finding perceptually relevant groups of sounds, as discussed by Wang and Bilger (1973). Until recently this grouping problem has remained unsolved. A solution, developed while writing this monograph, has now been proposed (Allen, 2005).

2.4.3 Event Classifications

The labels on the features of Fig. 2.12 are production based (i.e., Voicing, Nasality, Affrication, etc.), while in fact the features are based entirely on perceptual results. Would it not be better to initially ignore the production based labels and make them abstract quantities? While the phone classes which form diagonal blocks in the AM are highly correlated with production quantities, in fact they are psychophysically derived features (i.e., events).

Giving the events production based names is misleading, if not down right confusing. For example, what does the label "Voicing" really mean in this context? It is some derived quantity that psychophysically covaries with what we think of as voicing. In fact we do not actually know how the perceptual term "voicing" should be defined (Sinex *et al.*, 1991). A common definition is the time duration between the onset of the speech sound and the onset of glottal vibration. If this were the correct definition, then how could we decode voicing in the case of whispered speech? This problem has been explored at the auditory nerve level by Stevens and Wickesberg (1999, 2002).

Furthermore the MN55 articulatory feature classification scheme shown in Fig. 2.12 is seriously flawed. For example, the NASAL group are VOICED in the same sense as those labeled VOICED; however, the two clearly form a distinct clusters. There is no obvious simple articulatory label for sounds 8–14.

Groups systematically depend on the SNR, and groups remain unidentified by this scheme. Using the example of Fig. 2.11 (MN55 Table III), [/ba/, /va/, /ða/] form a group that is distinct from the nonfricative voiced subgroup. An improved order for sounds 8–14 would be [/ba/, /va/, /ða/], [/za/, /ʒa/, /da/, /ga/]. This example fundamentally breaks the MN55 articulatory feature classification scheme. In fact, it is likely that the feature space cannot be strictly articulatory based.

The events identified by Miller and Nicely are appropriate for the simple set of CVs that they studied, but are hardly inclusive. With other sounds, and for sounds from other languages, many other features will be necessary. Vowels have not been considered in any of these studies. The studies discussed in this review only represent a significant tip of a very large iceberg.

The data in the CM represents a psychological subject response, and therefore needs to be represented in terms of *psychological variables* rather than physical (production) measures, as defined by articulatory features. This could have been the role of *distinctive features*, had they been so defined. Unfortunately there seems to be some confusions in the literature as to the precise definition of a distinctive feature. For example, are distinctive features production or perception quantities?

To avoid this confusion, the term *event* is used when referring to *perceptual features*. Since Miller and Nicely's confusion data are based on perception, they must described events. The precise nature of these events may be explored by studying the 15 plots $S_{i,j}, i \neq j$, as shown in the lower-left panel of Fig. 2.13 for the case of $i = 2$.

2.4.4 The Transformation from CM to AM

In this section we transform the CM into an articulation matrix (AM), symmetrize the AM, and then re-express the SNR in terms of an *AI*. These transformations allow us to model the AM, giving incite into the detailed nature of the hierarchical categorical discriminations between the consonants. Much of the material from this section was extracted and independently published (Allen, 2005). While the two presentations are similar, they are not the same.

When normalized as a probability, the consonant CM is transformed to an *articulation matrix* (AM), denoted $\mathcal{A}$ (script-A), with elements

$$\mathcal{A}_{s,h} \equiv \frac{C_{s,h}}{\sum_h C_{s,h}}. \tag{2.25}$$

This normalization, to an equal probability for each row, is justified because of the small standard deviation of the row sums (i.e., 250 ± 21).

The AM is the empirical conditional probability $P_c(h|s)$ of reporting sound h after speaking sound s, namely

$$\mathcal{A}_{s,h} \equiv P_c(h|s) \tag{2.26}$$

for integer labels s, h (i.e., spoken, heard). In some sense $\mathcal{A}_{s,h}$ for $s \neq h$ is an error probability, since it is the probability of reporting the wrong sounds h after hearing spoken sound $s \neq h$.

Fig. 2.13 shows the probability of responding that the sound $h = 1, \ldots, 16$ was reported, following speaking /ta/ ($s = 2$), as a function of the wideband SNR. The upper-left panel shows the probability $\mathcal{A}_{2,h}(SNR)$ of each heard sound ($h = 1, \ldots, 16$), given /ta/ was spoken. The upper-right panel shows the

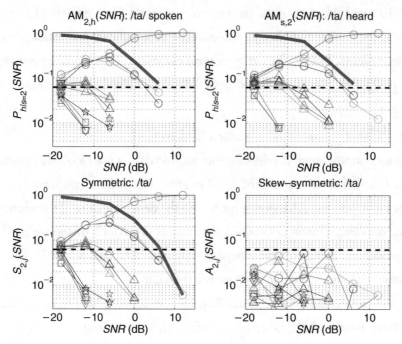

FIGURE 2.13: This figure shows Miller and Nicely's 1955 wideband row-normalized confusion matrix data $A_{s,h}(SNR)$ [Eq. (2.25)] for the sound /ta/ (sound 2) from MN55 Tables I–IV, as a function of the SNR. The upper-left panel is a plot of the second row of the articulation matrix [$A_{2,h}(SNR)$, $h = 1, \ldots, 16$, corresponding to /ta/ spoken], while the upper-right panel is a plot of the second column [$A_{s,2}(SNR)$, corresponding to /ta/ heard]. The matrix is not perfectly symmetric ($A \neq A^t$), which explains the small differences between these two plots. The lower-left panel is the symmetric form of the articulation matrix given by Eq. (2.27), which is the average of A and its transpose A^t. The lower-right panel is the skew-symmetric form $\check{A}$ [Eq. (2.28)]. The horizontal dashed line in each figure shows chance performance (i.e., 1/16).

probability $A_{s,2}$ of each sound spoken ($s = 1, \ldots, 16$), given that /ta/ was heard. The curve that rises to 1 is the probability of correctly reporting /ta/ ($A_{2,2}(SNR)$), given that it was spoken (left), or heard (right). The solid-thick curve is the total probability of error $e_2(SNR) \equiv 1 - A_{2,2}(SNR)$ of not reporting /ta/, given that it was spoken (left) or heard (right).

Symmetric and skew-symmetric decompositions: The lower-left panel of Fig. 2.13 is a plot of the second row $S_{2,j}$ of the symmetric form of the AM, defined as

$$S \equiv \frac{1}{2}\left(A + A^t\right),\tag{2.27}$$

where A^t is the transpose of A, while the lower-right panel is the second row of $A_{i,j}$ of the *skew–symmetric* form of the matrix, defined as

$$A \equiv \frac{1}{2}\left(A - A^t\right).\tag{2.28}$$

It appears that the sampling error (statistical uncertainty) in the measurements due to the sample size is about 0.5% (0.005), which is where the measurements become scattered. This variability is determined by many factors, including the number of trials per sound, the smoothing provided by the symmetric transformation, the consistency of the talker, and the mental concentration and number of the observers (four in this case).

From the lower-right panel, it is clear that the AM is close to symmetric, since the skew-symmetric terms are small. A few terms of $A_{2,h}(SNR)$ are as large as 5%, but most are less than 1%. Since the MN55 data are close to symmetric, it is reasonable to force the symmetry, and then to study S and A separately, which is the approach taken here. Note that S is slightly smoother than A, since each element $A_{s,h}$ is the average of two similar terms, $A_{h,s}$ and $A_{s,h}$. Using the symmetric form simplifies the analysis of the matrix and gives us access to the skew-symmetric form.

The interpretation of the skew-symmetric form is quite different from that of the symmetric form, The most likely explanation of the skew-symmetric matrix is that the subjects have a bias for one sound over another, and are therefore more likely to report the consonant for which they have the bias (Goldstein, 1980).[12]

Plotting the symmetric data $S(SNR)$ as a function of SNR, as shown in Fig. 2.13, provides a concise yet comprehensive summary of the entire set of measurements, and shows the hierarchical grouping, without a need to order the sounds. In the next section it is shown that if $S_{i,j}$ is described as a function of the AI, rather

---

[12] We have recently discovered that bias is not the explanation.

than the SNR, the same data may be quantitatively modeled, and the important effects of chance may be accounted for.

How can one make order (i.e., gain understanding) of the CM grouping, as shown in Fig. 2.11? The confusion matrix may be studied by forming a cascade of transformations, with the goal of studying the natural structure in the error patterns. Three transformations are used. The first transformation (T1) is to find the AI from the SNR. The second (T2) is to normalize the CM so that its row sums are 1. The third transformation (T3) is to express the normalized matrix as the sum of symmetric and skew-symmetric parts.

- **T1 (Miller Nicely and the AI):** To relate the Miller Nicely data to the Fletcher AI, $A(SNR)$ was computed, as follows. Since the ratio of the wide-band VU-level of the speech and of the noise are know (-18 to $+12$ dB), given the RMS spectrum of the speech (see Fig. 2.15) and of the noise (uniform from 0.1 to 9.5 kHz), the spectral levels in articulation bands may be computed, as shown in Fig. 2.16. From the ratio of these two articulation band spectra, $A(SNR)$ is easily found, from Eq. (2.13). The resulting $A(SNR)$ was found to vary from 0 to 0.6 as the SNR is varied from -18 to $+12$ dB.

- **T2 (Row normalization):** The *second* transformation is to normalize the rows of the confusion (or Count) matrix $C_{r,c}(SNR)$ by the row-sum, thereby converting the confusion matrix into an empirical proba-bility measure $A_{r,c}(SNR)$ [i.e., an *articulation matrix* (AM)] as show by Eq. (2.25). The resulting $A_{s,h}(SNR)$ is a matrix of PI functions, with row index $s = 1, 2, \ldots, 16$ indexing the stimuli spoken and column index $h = 1, 2, \ldots, 16$ indexing the responses heard.

- **T3 (Decomposition):** A *third* transformation is to form the symmetric[13] and skew symmetric components from matrix $A(SNR)$. The symmetric

---

[13] The two transformations, normalization and symmetry, interact, so it is necessary to iterate this pair of transformations to convergence. For the tables provided by Miller Nicely, this iteration converges rapidly, as the matrix is close to symmetric to start.

form of the AM is given by Eq. (2.27), while the skew-symmetric form is defined by Eq. (2.28).

There is an interaction between the row normalization [Eq. (2.25)], and the symmetry transformation Eq. (2.27), which requires that the row-normalization and symmetric computations be iterated. This iteration always converges to the same result, and is always stable for all of the MN55 tables. An entry of "1" in C represents a single vote for the same utterance, from 4 listeners who heard that utterance. All 1's were deleted from the matrix before computing S, which made the functions much smoother as a function of SNR at very low probabilities. Once matrix S has been determined, A is computed from

$$A = A - S. \tag{2.29}$$

2.4.5 Hierarchical Clusters

Sound clustering in the CM was used by MN55 as the basis for arguing that the sounds break down into distinct groups, which MN55 identified as five discrete *articulatory features*, and called these as *Voicing, Nasality, Affrication, Duration*, and *Place*.

Each symbol in Fig. 2.13 labels a different articulatory feature. Sounds 1–3 (/pa/, /ta/, /ka/) are shown as circles, 4–7 (/fa/, /θa/, /sa/, /ʃa/) triangles, 8–10 (/ba/, /da/, /ga/) squares, 11–14 (/va/, /ða/, /za/, /ʒa/) upside-down triangles, while the nasal sounds 14 and 15 (/ma/, /na/) are labeled by 5-pointed stars.

The hierarchical clusters are seen in $S_{r,c}(SNR)$ as groups that peel away as the SNR (or AI in the case of Fig. 2.21) increases. The symmetric /ta/ data shown in the lower-left panel of Fig. 2.13, is a great example: First all the voiced sounds dramatically drop, starting from chance, as the SNR is raised. Next the unvoiced-fricatives /fa/, /θa/, /sa/, and /ʃa/ (triangles) peel off, after very slightly rising above chance at −12 dB SNR. Finally the two main competitors to /ta/ (/pa/ and /ka/) peak around −6 dB SNR, and then fall dramatically, as /ta/ is clearly identified at 0 dB SNR and above. In the lower-left panel /pa/, /ta/, and /ka/ (◯) are statistically indistinguishable below −6 dB and approach chance identification of 1/16 at −18 dB. Above about −6 dB, /ta/ separates and the identification approaches 1, while the confusions with the other two sounds (/pa/ and /ka/) reach a maximum of about a 25% score, and then drop monotonically, as the SNR increases.

The MN55 sounds 4–7 (/fa/, /θa/, /sa/, and /ʃa/), like sounds 1–3 (/pa/, /ta/, /ka/), also form a group, as may be seen in the lower-left panel, labeled by △. This group also starts from chance identification (6.25%), rises slightly to a score of about 7%, at −12dB, and then monotonically drops at a slightly greater rate than sounds 1 and 3 (symbols ○).

The remaining sounds 8–16, labeled by the remaining symbols, which show no rise in performance; rather they steeply drop, from the chance level.

At the lowest SNR of −18 dB, the elements in the symmetric form of the AM approach chance performance, which for MN55 is 1/16, corresponding to closed-set guessing. Extrapolating the data of Fig. 2.13, chance performance corresponds to about −21 dB SNR.

Based on the clustering seen in the AM (e.g., MN55 Tables II and III), it was concluded by MN55 that the three sounds /ta/, /pa/, and /ka/ might be thought of as one group. These three sounds form the unvoiced, nonnasal, nonaffricate, low-duration group, having three different values of place. The details of these groupings depend on the SNR. A detailed analysis of these clusters show that the MN55 *articulatory features* (production feature set) do not always correspond to the *events* (perceptual feature set).

In fact it would be surprising if it turned out any other way, given that production and perception are fundamentally different things. The details of a scheme that will allow us to make such an analysis of the optimal perceptual feature set, forms the remainder of this section.

2.4.6 Total Error Decomposition

The solid-thick curve in the top two and bottom-left panels of Fig. 2.13 are graphs of the total error for /ta/

$$e_2(SNR) \equiv 1 - S_{2,2}(SNR). \tag{2.30}$$

Because each row of $S_{i,j}$ has been normalized so that it sums to 1, the total error for the ith sound is also the row sum of the 15 off-diagonal ($j \neq i$) elements, namely

$$e_i(SNR) = \sum_{\forall j \neq i} S_{i,j}(SNR). \tag{2.31}$$

Since each error term is nonnegative, e_i must bound each individual confusion $S_{i,j}$. For the data of Fig. 2.13, lower-left, the other two circle curves (/pa/ and /ka/), which compete with /ta/, and thereby form a three-group, are nearly identical. All other error terms are much smaller. Thus the solid-thick curve, $e_2(SNR)$, is approximately twice the size of the curves for /pa/ and /ka/. All the off diagonal terms go to zero at $+12$ dB SNR so for that one point $e_2 = S_{2,j}$, a fluke of the small-number statistics.

Equation (2.31) will turn out to be a key decomposition that allows us to break the total error down into its parts. The total error for the ith sound is linearly decomposed by the off-diagonal errors of the AM. The sounds that are confusable have larger error, and the sounds from different groups contribute, down to the chance level. This is a natural decomposition of the total error into its confusions, that can help us understand the AI predictions in much greater detail.

For example, *why does the probability of identification of sounds 1–3 and 4–7 increase even when these sounds are not spoken?* The initial rise for the two sound groups follows from the increase in chance performance due to the decreased entropy, which follows from the reduced size of the group. This conclusion follows naturally from Eq. (2.31). As the SNR increases, the size of the group exponentially decreases.

As the number of alternatives in a closed-set task decreases, the probability of guessing increases. Given 2 alternatives, chance is 1/2; given 16, chance is 1/16. Thus grouping and the rise due to the confusion within the group, are intimately tied together. In the same manner, as the SNR rises from -18 to -12, the MN55 sounds 4–16 are perceptually ruled out, increasing chance performance for sounds 1–3 from 1/16 to 1/3.

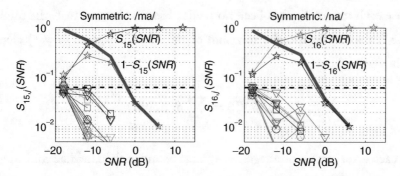

FIGURE 2.14: Plots of the symmetric AM corresponding to the nasals /ma/ and /na/. The curve that rises to 1 is $S_{i,i}(SNR)$ for $i = 15$ (left) and $i = 16$ (right). The solid fat curve in each panel is e_i [Eq. (2.31)]. The other curves represent confusions $S_{i,j}(SNR)$ for the remaining sounds $j = 1, \ldots, 14$.

It seems obvious that the reduced entropy with SNR results from events (perceptual features) common to each group. No better example of this effect is the case of the nasals, which form a group of 2 in the MN55 data set.

The nasals: In Fig. 2.14 $S_{i,j}(SNR)$ for $i = 15, 16$, corresponding to /ma/ and /na/, are presented. The two nasal sounds are clearly separated from all the other sounds, even at -18 dB SNR. As the SNR increases, the scores rise to $\approx 25\%$, peaking at or near -12 dB SNR, following with the identification rising and the confusion dramatically falling for SNRs at and above -6 dB.

Sounds 1–14 are solidly rejected, even at -18 dB. These scores exponentially drop as the SNR is increased. There is a slight (visual) hint of a rise of a few sounds for the case of /ma/, in some of the rejected sounds in the left panel, and some corresponding grouping, but the effect is small and it would be difficult to tease out. The rejected sounds in the right panel do not show any obvious grouping effect.

The subjects can clearly distinguish the two nasal sounds (sounds 15,16) from all the others (sounds 1–14), even at the lowest SNR of -18 dB; however, they cannot distinguish between them until the SNR is greater than -12 dB. The subjects know the sound they hear is NASAL, but the question is, which one? This identification of event-NASAL leads to a significant increase in chance performance for SNRs between -18 and -6 dB, from $1/16$ to $1/2$.

One may also see this effect in the raw count data at -18 dB, where the confusions are approaching equal chance levels. For example, in MN55 Table I, the raw counts are [25, 28; 33, 32]. At -12 dB, /ma/ and /na/ are significantly confused with each other, but rarely with the other sounds. For example, from MN55 Table II, /ma/ is heard 20 times when /ba/ is spoken, ($S_{15,8}(-12) = 6.72\%$ of the time), while /ba/ is heard 11 times when /ma/ is spoken (5.83% of the time).

2.5 TRANSFORMATION FROM THE WIDEBAND SNR TO THE AI

Miller and Nicely used the wideband SNR, in dB, as their measure of audibility. However, as discussed in the introduction, there are reasons to believe that the AI(SNR) is a better audibility measure. We shall now demonstrate this for the MN55 data. Our approach is to transform MN55's wideband SNR into an *AI* and then to plot the resulting $S_{i,j}(AI)$.

To compute the AI for MN55 one needs to know the *specific SNR*, over articulation bands, denoted SNR_k. This requires knowledge of the average speech spectra for five female talkers, and the noise spectra. The spectrum for five female talkers is shown in Fig. 2.15, while the noise spectra was independent of frequency (i.e., white). The procedure for computing *AI(SNR)* is described next.

2.5.1 Computing the Specific AI

The AI is defined by French and Steinberg (1947, Eq. 8) as Eq. (2.13), namely as a 20 band average over the *specific AI*, denoted AI_k. The specific AI is defined in terms of the SNR

$$SNR_k \equiv \sigma_{s,k}/\sigma_{n,k}. \tag{2.32}$$

where the speech power is $\sigma_{s,k}^2$ [Watts/critical-band] and the masking noise power is $\sigma_{n,k}^2$ [Watts/critical-band], in the kth articulation band. When calculating $\sigma_{s,k}$,

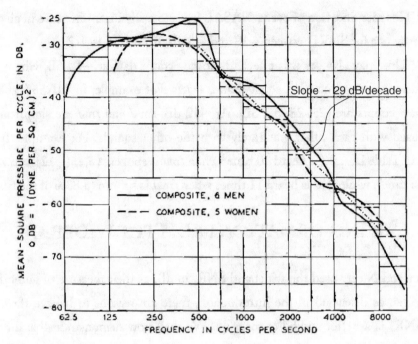

FIGURE 2.15: This figure from Dunn and White (1940, Fig. 10) shows the average power spectrum for 6 men and 5 women. The dashed curve, which approximates the power spectrum for the 5 women, has a slope of 0 from 125 to 500 Hz, and a slope of −29 dB/decade between 0.5 and 8 kHz.

the average is over 1/8-sc intervals. SNR_k is the same as French and Steinberg's *band sensation level*, which they denoted as E. The kth articulation band power-snr *speech detection threshold* may be modeled as

$$\frac{I + \Delta I}{I} \equiv \frac{\sigma_{n,k}^2 + c^2 \sigma_{s,k}^2}{\sigma_{n,k}^2} = 1 + c^2 snr_k^2, \qquad (2.33)$$

where a frequency independent *speech detection constant* c is determined empirically from data on the detection of speech in noise (Fletcher and Munson, 1937; French and Steinberg, 1947). The role of c is to convert the speech RMS to the speech peaks, which are typically 12 dB above the RMS speech level. When SNR_k is specified in terms of speech peaks, $c = 2$.

Converting to decibels, and scaling by 30, defines the *specific AI*

$$AI_k = \min\left(\frac{1}{3}\log_{10}\left(1 + c^2 snr_k^2\right), 1\right). \qquad (2.34)$$

Relationship Eq. (2.34) follows from the detailed discussions of French and Steinberg (1947) and Fletcher and Galt (1950), followed by the subsequent analysis by Allen (1994) [see especially, Fletcher 1995, Eq. (10.3), p. 167], and is more accurate than Eq. (2.16) in the neighborhood of the speech detection threshold.

Between 0 and 30 dB, AI_k is proportional to $\log(SNR_k)$ because the percent of the time the speech is above a certain level is proportional to the dB-SL level (rethreshold sensation level) (French and Steinberg, 1947; Allen, 1994). The factor of 1/3 comes from the dynamic range of speech (30 dB) which is used as the normalization factor in a given articulation band (French and Steinberg, 1947, Fig. 4, p. 95). As discussed extensively by French and Steinberg (1947, Table 12), an empirical *threshold adjustment* must be made, labeled c in Eq. (2.33). The value of c is chosen such that the speech is just detectable when $cSNR_k = 1$, in each cochlear critical band, corresponding to specific AIs of zero (i.e., $AI_k = 0$). In the present computations this adjustment was 6 dB ($c = 2$), as was empirically determined by French and Steinberg (1947, Eq. 12, Fig. 21). A more precise estimation of c will require repeating Fletcher's critical ratio experiment using narrow bands of speech, with a white noise masker and measuring SNR_k at the detection threshold. The $\min(x, 1)$ part of the definition limits the AI on the high end, since for an SNR above 30 dB, the noise has a negligible effect on the articulation (and intelligibility).

2.5.2   The Band Independence Model of the Total Error

The average sound *articulation error* $e(SNR)$, in terms of the average sound articulation $P_c(SNR)$, is

$$e(SNR) = 1 - P_c(SNR). \qquad (2.35)$$

In 1921 Fletcher showed that the articulation error probability $e(SNR)$ could be thought of as being distributed over K independent articulation bands. The bandwidth of each of these articulation bands was chosen so that they contribute equally to e (the articulation per critical band is constant from 0.3–7 kHz (Fletcher and Galt, 1950; Allen, 1994, 1996). Assuming band independence, the total articulation error may be written as a product over K band articulation errors, as given by Eq. (2.12).

R. Galt established that the articulation bandwidth is proportional to cochlear critical bandwidths (French and Steinberg, 1947, p. 93), as measured by the *critical ratio* method and the frequency JND (Allen, 1994, 1996). Fletcher then estimated that each articulation band was the equivalent of 1 mm of distance along the basilar membrane, thereby taking up the 20 mm distance along the basilar membrane, between 300 to 8 kHz (Allen, 1996). Thus the AI [Eq. (2.13)] may be viewed as an average SNR, *averaged over dB units*, of a scaled specific SNR, defined over cochlear critical bands.

As first derived in Allen (1994), the probability of articulation error in the kth band ϵ_k may be written in terms of the specific AI as

$$\epsilon_k = e_{\min}{}^{AI_k/K}, \tag{2.36}$$

where the constant $e_{\min}$ is defined as the minimum error via the relationship

$$e_{\min} \equiv 1 - \max_{SNR}\left(P_c(SNR)\right). \tag{2.37}$$

This constant $e_{\min}$ depends in general on the corpus, talkers and subjects. For Fletcher's work, $e_{\min}$ was 1.5% ($\mathcal{H} \approx 11$, i.e., more than 2048 sounds). For the work reported here, a value of 0.254% ($\mathcal{H} = 4$) was used, based on an extrapolation of the MN55 data to $AI = 1$ and a minimization of the model parameters $e_{\min}$ and c for a best fit to the MN55 data.

Equation (2.10) follows from the above relations, and only applies to the case of MaxEnt phones, such that the phone entropy is maximum.

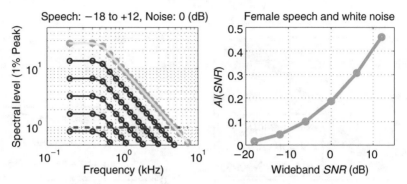

FIGURE 2.16: Using the speech power spectrum given by the dashed line in Fig. 2.15, and assuming a uniform noise spectral level, the $AI(SNR)$ was calculated. Each curve shows the relative spectral level of the speech having a peak RMS level at the wideband SNRs used by Miller and Nicely [−18, −12, −6, 0, 6, 12], in units of dB. The top curve shows the +12 dB speech spectrum. The dashed-dot line is the noise spectral level having an RMS of 0 dB.

Fig. 2.16, left, shows the relative spectrum and noise level corresponding to SNR's of −18 to +12 dB, for female speech with a white noise masker. On the right one may see the resulting $AI(SNR)$, based on the calculations specified by the equations presented in this section. The final values of the AI were determined with $c = 2$ to be (starting from an SNR of +12): [0.459, 0.306, 0.186, 0.1, 0.045, 0.016].

Because the spectrum of the speech and the spectrum of the noise are not the same, the AI(SNR) cannot be a linear function of SNR. Only for the case where the two spectra have the same shape, will AI(SNR) be linear in SNR. For the case in hand, a white noise masker, the high frequencies are progressively removed as the SNR decreases, as shown in the left panel of Fig. 2.16.

2.5.3 AM(SNR) to AM(AI)

The left panel of Fig. 2.17 shows the MN55 consonant identification curves $P_c^{(i)}(SNR) \equiv S_{i,i}(SNR)$, as a function of the SNR for each of the 16 sounds $(i = 1, \ldots, 16)$, along with their mean $P_c(SNR)$ (solid curve with circle symbols)

$$P_c \equiv \frac{1}{16} \sum_{i=1}^{16} P_c^{(i)}. \tag{2.38}$$

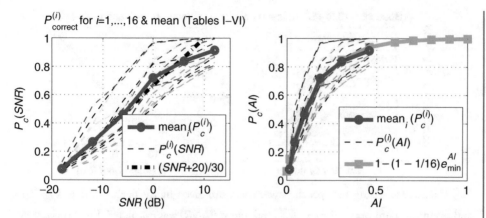

FIGURE 2.17: The light dashed lines are $P_c^{(i)}$ for each of the 16 consonants. On the left the abscissa is the SNR in dB, while on the right, the AI is used as the independent variable. The solid-thick curve (circles) on both the left and right is the average score P_c Eq. (2.38). The solid-thick curve (squares) on the right is the average phone prediction given by Eq. 2.39.

It must be mentioned that Eq. (2.38) applies only to the case at hand, where the *a priori* probabilities of the sounds are equal (i.e., 1/16). In the more general case, a Bayesian formulation would be required.

In the right panel the individual scores, along with the average, are shown as a function of the AI. To transform from SNR to AI the values shown in the right panel of Fig. 2.16 are used as determined by a sum over terms defined by Eq. 2.34 with $c = 2$.

We also wish to compare the AI model prediction to the measurements shown in Fig. 2.17. However it is necessary to modify Eq. (2.10) so that it accounts for chance (guessing) given by $P_{\text{chance}} = 2^{-\mathcal{H}}$, when $\mathcal{H} = 4$ and $AI = 0$. This is done by again assuming independence of the error probabilities. Since chance error for guessing is $e_{\text{chance}} = 1 - P_{\text{chance}}$, the chance-corrected $P_c(AI)$ formula is

$$P_c(AI, \mathcal{H}) = 1 - e_{\text{chance}}(\mathcal{H}) \; e_{\min}^{AI} \tag{2.39}$$

with

$$e_{\text{chance}}(\mathcal{H}) \equiv 1 - 2^{-\mathcal{H}}. \tag{2.40}$$

Fletcher's formula Eq. (2.10) is the limiting case of Eq. 2.39 when $\mathcal{H}$ becomes large (Fletcher's $\mathcal{H} \approx 11$).

A plot of Eq. 2.39 is shown in the right panel of Fig. 2.17 (solid curve, square symbols), with $e_{\min} = 0.254\%$, and $\mathcal{H} = 4$. The fit of Eq. 2.39 to the average of the 16 MN55 curves is excellent.

Discussion: The left panel of Fig. 2.17 shows that there is an approximately linear relationship between $P_c(SNR)$ and SNR over the range from -18 to 6 dB. The thick dashed-dot line is (SNR+20)/30. This line is useful as a simple reference.

The main deviation from the linear dash-dot curve is due to the strong saturation that occurs for the two nasal sounds and sound 7 (the three curves with the highest $P_c(SNR)$). Note that each of the sounds have a nearly linear $P_c^{(i)}(SNR)$, with different saturation levels (if they are reached). The saturation point for $P_c(SNR)$ occurs at an SNR of about 30 dB above the threshold, at -20 dB (fat solid line with circles). Note that since $P_c(SNR)$ depends on the noise spectrum, the linear relation observed in the left panel of Fig. 2.17 can only hold for the white noise masker, since if the noise spectrum is changed, $P_c(SNR)$ must change, and it is linear for the white noise case.

In the right panel of Fig. 2.17 the extended AI model (Eq. 2.39) is shown for MN55's data. Each of the 16 curves $P_c^{(i)}(AI)$, $i = 1 \ldots 16$, are shown as the light-dashed curves. This average [Eq. (2.38)] is shown as the solid-thick curve with circles.

The solid-thick line with squares is the extended (chance-corrected) AI model, Eq. 2.39. The value of $e_{\min}$ of 0.254% is one sixth that used by Fletcher (1.5%). The smaller size could be attributed to the larger amount of training the subjects received over such a limited set size $\mathcal{H} = 4 = \log_2(16)$.

As may be seen in the left panel of Fig. 2.16, since MN55 used white noise, the SNR_k for frequency bands larger than about 0.7 kHz have an SNR of less than 30 dB, resulting in an AI of much less than 1. In fact the AI was less than than 0.5 for the MN55 experiment, corresponding to a maximum score of only 90%.

A most interesting (and surprising) finding is that the extended AI model (Eq. 2.39) does a good job of fitting the average data. In fact, the accuracy of the fit over such a small set of just 16 consonants was totally unanticipated. This needs further elucidation.

2.5.4   Extended Tests of the AI Model

If one plots the total error probability $e(AI) = 1 - P_c(AI)$ on log coordinates, as a function of AI, such plots should approximate straight lines. This follows from the log of Eq. 2.39:

$$\log(e(AI)) = \log(e_{\min}) AI + \log(e_{\text{chance}}(\mathcal{H})), \qquad (2.41)$$

which has the convenient form $y = ax + b$. The ordinate (y-axis) intercept of these curves at AI = 0 gives the log chance-error [$b \equiv y(0) = \log(e(0)) = \log(e_{\text{chance}}(\mathcal{H}))$], while the ordinate intercept of these curves at $AI = 1$ defines the sum of the log-chance error and the log-minimum error, namely [$a + b \equiv y(1)$ thus $a = \log(e_{\min})$]. In Fig. 2.18 the log-error probabilities for each of the 16 sounds, along with the average and the AI model, are shown. The sounds have been regrouped so that the log-error plots have similar shapes. The shallow slopes are shown on the left and the steeper slopes on the right.

From Fig. 2.18, we will see that the linear relationship [Eq. (2.41)] holds for 11 of the 16 sounds, with the free parameters $e_{\min}(i, j)$ and $e_{\text{chance}}(i)$, either depending on the sound, or on a sound group.

The upper two panels show the most linear groups, while the lower panels are the most nonlinear (non-straight) log-error curves. The curves that are close to linear (the two top panels) are consistent with the AI model, due to Eq. (2.41).

This observation of a log-linearity dependence for the probability of error of individual sounds is rather astounding in my view. First, there was no *a priori* basis for anticipating that individual sounds might obey Fletcher's band-independence property, Eq. (2.9). Second, if individual sounds obey equations of the form of

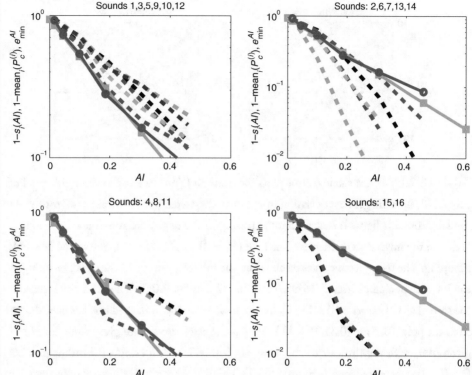

FIGURE 2.18: This figure shows the probability of error for the ith sound, $P_e^{(i)}(AI) \equiv 1 - P_c^{(i)}(AI)$, as a dashed curve. To reduce the clutter, the sounds have been sorted over the four panels, with the sound number indicated in each panel title. The top two panels are the cases where the individual sound error curves are close to straight lines. The left-upper panel are those cases where the sound lies above the average, while the right-upper panel shows those cases where the sound lies below the average. The two lower panels correspond to the sounds that violate the exponential rule (are not straight lines on a log-error plot). For reference, each panel contains the average probability of error $P_e(AI) \equiv 1 - P_c(AI)$, shown as the solid curve with circles, and the model error $e_{\min}{}^{AI}$, shown as the solid line (squares).

Eq. (2.39), then sums of such equations cannot obey Eq. (2.39), since the sum of many exponentials, each having a different base, is not an exponential.

The finding that individual CV recognition error is exponential in the AI (the basis of the band independence hypothesis) therefore extends, and at the same time violates, Fletcher's original fundamental AI hypothesis that the average error is exponential.

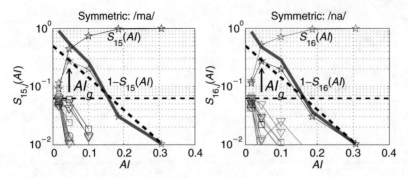

FIGURE 2.19: Since the log-error plots for /ma/ and /na/ (see the lower-right panel of Fig. 2.18) show the greatest deviation from linear, they seem to be a "worst case" for the AI model. From this figure it is clear that the reason for the deviation from linear dependence is due to the migration of chance from 1/16 ($\mathcal{H} = 4$) to 1/2 ($\mathcal{H} = 1$), due to the NASAL grouping. The rising nasal curves results from the robust grouping of the nasal, resulting in the increase in chance from 1/16 at $AI = 0$ to 1/2 at $AI \approx 0.045$. The solid-thick curve is the sum of all the errors (and is $1 - P_e$ for the spoken sound). A dashed line has been drawn from the point (0,.5) to (0.31,.01). This line fits the error curve (/na/ given /ma/, and /ma/ given /na/) with very little error, for $AI = AI_g > 0.045$, and intercepts the ordinate at 1/2 for $AI = 0$, as expected for a 2-group ($\mathcal{H} = 1$). This further supports the band independence model Eq. (2.12).

It is therefore essential to understand the source of the deviations for the individual sounds from the average, and to critically assess the accuracy of the model for individual sounds. Five sounds (4, 8, 11, 15, 16) have a probability of error that deviates from linear, with the most nonlinear and the largest deviations from the mean, being the nasals (15,16), as shown in the lower-right panel. In the next section we explore the reasons for this.

Log-error for the nasals: In Fig. 2.19 the nasal data are shown using the same log-error linear decomposition used in Fig. 2.14, where the total error (solid-thick curve) is the sum of the errors of the competing sounds [i.e., Eq. (2.31)]. In the case of the nasals, the confusions for the other sounds is small, namely only /ma/ and /na/ significantly compete.

As a result of plotting the data as a function of AI, for $AI > AI_g = 0.045$ (SNR $\geq$-12), the log-error curves become linear in AI, as predicted (modeled) by Eq. 2.39. This value of AI_g is shown in the plot with an arrow indicating the point of separation of the target sound from the competing sound. Extrapolation of this linear region back to $AI = 0$, one finds the chance guessing probability of $1 - 2^{-\mathcal{H}_g} = 1/2$, corresponding to a nasal group entropy of $\mathcal{H}_g = 1$. This is shown on the graph by the dashed line superimposed on the corresponding error curve (stars). In the region $0 \leq AI \leq AI_g = 0.045$, $\mathcal{H}$ depends on AI, since it dramatically drops from 4 to 1.

Thus the reason that the nasal curves are not linear in Fig. 2.18 is that chance (the entropy factor) is dramatically changing between $0 \leq AI \leq AI_g$, due to the formation of the perceptual "event-nasal" group.

When the data is plotted as a function of SNR, as in Fig. 2.14, the log-error linearity is not observed. Also the shape of the curve will depend on the spectrum of the noise. Clearly the SNR to AI transformation is an important key to making sense of this data.

One may assign each sound to a unique group by grouping all the sounds having off-diagonal PI functions that initially rises from chance, as a function of the SNR. The strength of the group may be characterized by the value of the SNR where the functions have a local maximum ($SNR_g = -12$ dB in the example of Fig. 2.14 or $AI_g = 0.045$ for the data of Fig. 2.19).

Log-error for /pa/, /ta/, and /ka/:   Finally in Fig. 2.20 we return to the case of /pa/, /ta/, and /ka/. This three-group generalizes the /ma/, /na/ two-group conclusions of Fig. 2.19. In the middle panel it is clear that for small values of AI less than 0.045 $S_{2.2}(AI)$ for /ta/ is equal to the curves for /pa/ and /ka/ ($S_{2.j}(AI)$, $j = 1, 3$). As the AI rises above about 0.1, the three curves (circles) split due to the identification of /ta/ and the rejection of /pa/ and /ka/. The shape and slope of the curves corresponding to the two rejected sounds are identical. The projection of the rejected curves back to $AI = 0$ gives the probability of chance-error for a group of 3 (i.e., 1-1/3), as shown by the dashed line in this middle panel. In the left-most and right-most panels, corresponding to /pa/ and /ka/, the two rejected sounds have very different log-error slopes. However, the two dashed curves still project back

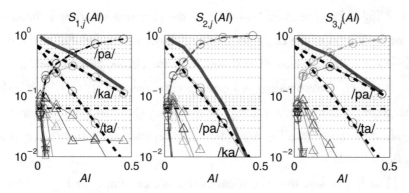

FIGURE 2.20: This figure shows $S_{s,b}(AI)$ for $s = 1, 2, 3$ corresponding to the sounds /pa/, /ta/, and /ka/. The dashed lines connect (0,1-1/3) with (0.48, 0.1) and (0.442, 0.01). A second weak group ($AI_g \approx .01$) labeled by the triangles corresponds to the Affrications, sounds 4-7.

to the chance error probability for a group of 3 (1-1/3). This change in the slope for the two sounds shows that $e_{\min}(i, j)$ can, in general, depend on the sound in the group. This seems to reflect the more robust nature of /ta/ relative to /pa/ and /ka/ due to /ta/ having more high frequency energy than its competitors.

Based on the small amount of the data shown in Fig. 2.19 and Fig. 2.20, it appears that the band independence assumption Eq. (2.12) and the band error expression Eq. (2.36) model the individual sound confusions $S_{i,j}(AI)$ more accurately than they model the average band error [Eq. (2.12)]. The total sound error is more precisely the sum of these off diagonal confusion terms, as given by Eq. (2.31). The implications of this model seem quite significant, but without more data, it is unwise to speculate further at this time.

The symmetric AM: The upper half of Fig. 2.21 shows all the $S(\mathcal{A})$ data. The first panel (upper-left) shows $S_{1,r}(\mathcal{A})$, with $r = 1, 2, \ldots, 16$. The lower half of the figure shows the skew symmetric matrix A, as a function of the scaled *snr*, $(SNR + 20)/30$.

The skew symmetric AM: As shown in Fig. 2.21 a small number of the skew symmetric PI function-pairs are larger than the others. For example, only 6 have more than 5% error, and none are more than 10%. Thus the ordinate of the skew symmetric PI functions has been expanded by a factor of 10. For each skew symmetric PI function $A_{r,c}(\mathcal{A})$, there is a corresponding symmetric negative PI function in one of the neighboring panels. An analysis of these pairs identifies individual curves. For example, $/fa/$ and $/\delta a/$ are the largest matching curves of opposite

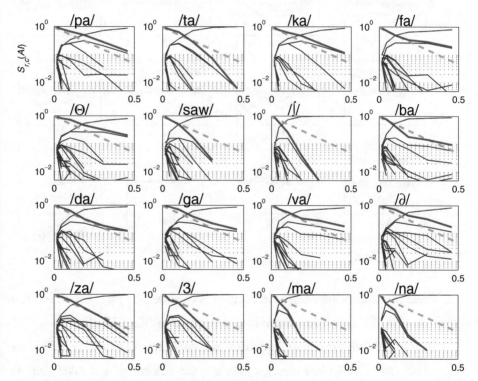

FIGURE 2.21: This chart presents the confusion Tables I–VI (the wideband data) $\mathcal{A}$ after being split into the symmetric and skew symmetric parts. The upper 4×4 display shows Eq. (2.27) the symmetric component $S(AI)$ of the AM. The thick-solid curve is the diagonal element $r = c$, while the off diagonal sounds are given by the remaining thin-solid curves. The dashed curve is Eq. 2.39. The lower panel shows Eq. (2.28), the skew symmetric component $A(SNR)$, as a function of the normalized signal to noise ratio $(SNR + 20)/32$ (Using the AI as the abscissa for the skew component does not make sense.). The ordinate of $A(SNR)$ has been scaled to ± 0.15 to magnify the scale of the naturally small skew component of the AM.

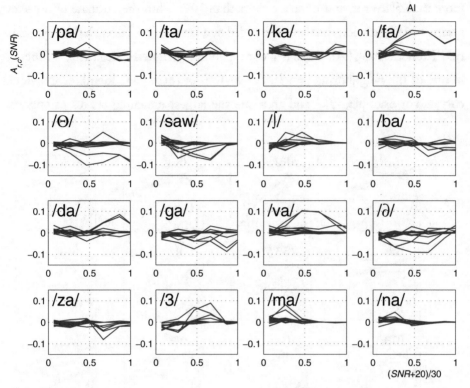

FIGURE 2.21: (Cont.)

sign. These two sounds are being asymmetrically confused (in place), with an error as large as 10%. This confusion extends to the largest *SNR*, of 12 dB.

The sounds with the most significant skewness are */va/–/θa/* (a 10% place error), */da/–/ga/* (a 9% place error), */ʒa/–/ga/* (an 8% error), and */ða/–/sa/* (a 7% place error). All the remaining asymmetric confusions have errors of 5% or less, which may be below the level of significance. Unlike the */ʃa/* and */ða/* case, these confusions become small as the *SNR* increases.

Would these confusions go to zero if the *SNR* were further increased? The problem here may be the bandwidth, not the *SNR*. As we shall see, frequencies above 6.5 kHz are missing in MN55, thus the AI never goes above 0.6, resulting in problems with sounds having high frequency articulations. The flat spectrum of the noise masks the high frequency components of the speech.

If these larger entries in $A_{r,c}$ are due to listener bias and subtle talker dialect imperfections, then it would be best to remove this component, as may be done by analyzing S rather than A. Two obvious questions are "Why is A so close to symmetric?" and "Is it *a priori* obvious that A should be symmetric?" Symmetric channels have important properties that might aid in our further data analysis, thus there is a large stake in answering the above questions.

2.5.5   The AI and the Channel Capacity

It is significant that the band average $\overline{AI_k}$ [Eq. (2.13)] is over in dB units, defined by Eq. (2.16), rather than over power (i.e., $\sum_k SNR_k^2$). Since the sum of logs is the log of products (i.e., $\log x + \log y = \log xy$):

$$\frac{1}{K} \sum_k SNR_k \propto \log \left(\prod_k snr_k \right)^{1/K}, \tag{2.42}$$

the *AI* may be defined in terms of a geometric mean of *SNRs*. This is a subtle and significant observation that seems to have been overlooked in discussions of the AI. The geometric mean of the snr_k over frequency is used in information theory as a measure an abstract volume, representing the amount of information that can be transmitted by a channel (Wozencraft and Jacobs, 1965). For example, the Shannon Gaussian channel capacity formula

$$C = \int_{-\infty}^{\infty} \log_2[1 + snr^2(f)]\, df, \tag{2.43}$$

which is a measure of a Gaussian channel's maximum capacity for carrying information, is conceptually very similar to Eq. (2.42).[14] From Fig. 2.22, we see that $AI_k(SNR_k)$ as given by Eq. 2.16 is a straight–line approximation to the integrand of

---

[14] A discussion with Leo Beranek at the 2002 Winter meeting of the ASA reveled that he was aware of the similarity between C and A as early as 1945, but the observation was never explored.

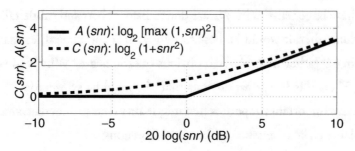

FIGURE 2.22: Plot of $\log(1 + SNR^2)$ and $\log[\max(1, SNR)^2]$ versus $SNR = 20 * \log(SNR)$.

the Shannon channel capacity formula $C(SNR)$. The figure shows the two functions $C(SNR) \equiv \log_2[1 + snr^2]$ and $A(SNR) \equiv \log_2[\max(1, snr)^2]$.

The first formulation of the channel capacity, as proposed by R. V. L. Hartley (Hartley, 1928; Wozencraft and Jacobs, 1965), was to count the number of intensity levels in units of noise variance. This is a concept related to counting JNDs in psychophysics. It is interesting and relevant that R. V. L. Hartley was the father of the decibel, which was also based on the intensity JND.[15] Hartley, a Rhode Scholar, was well versed in psychophysical concepts (Hartley and Fry, 1921). The expression

$$\log(1 + snr^2) = \log\left(\frac{P + N}{N}\right), \quad (2.44)$$

where P and N are the signal and noise powers respectively, is closely related to counting JNDs. It has been observed, by George A. Miller (Miller, 1947b), that a signal change is close to the first JND level if its presence changes the input stimulus by 1 dB, that is when

$$10\log_{10}\left(\frac{P + N}{N}\right) = 1. \quad (2.45)$$

---

[15] The JND is the *just noticeable difference*. The *intensity JND* is the just noticeable change in intensity (Fletcher, 1995; Allen and Neely, 1997). Counting JNDs has a long history in psychophysics. The idea of using the variance as a scale was introduced by Thurstone (Baird and Noma, 1978). The concept of counting JNDs is attributed to Fechner (circa 1860) (Fechner, 1966), who is viewed as the *Father of Psychophysics*. For example, see Ch. 2 of (Baird and Noma, 1978) at http://auditorymodels.org/jba/BOOKS_Historical/Baird/.

Hence, the function $\log(1 + c^2 SNR^2)$ in fact expresses the noise level in terms of the JND (French and Steinberg, 1947; Fletcher and Galt, 1950; Kryter, 1962b; Siebert, 1970; Allen and Neely, 1997).[16] The product of the number of articulation bands times the number of JNDs determines a volume, just as the channel capacity determines a volume.

2.6   SINGULAR VALUE DECOMPOSITIONS OF THE AM SYMMETRIC FORM

To transform a symmetric matrix into a block-symmetric form, one must know how to group (i.e., order) the sounds. This grouping is closely related to statistical clustering and scaling techniques. Multidimensional scaling (MDS) methods were one of the earliest transformations that were systematically explored, as reviewed in the classic reports of Shepard (1972, 1974) and Wang and Bilger (1973).

Other possible transformations exist. For example, one alternative scaling is a *permutation*. A permutation matrix $\mathcal{P}$ has the property that $\mathcal{P}^2 = 1$, namely the second application of any permutation, undoes its effect. An example of a permutation is the interchange of two rows and their corresponding columns. A more general linear transformation corresponds to a pure rotation about some point. The family of rotations are said to be *orthogonal* and satisfy the property $U^T U = 1$, where U^T is the *transpose* of U, corresponding to mapping the rows of U into the columns of U^T. Rotations are much more general than permutations, which make up a *subgroup* of the rotations. The AM would have a totally different form after any such orthogonal rotations. The question is, what general type of linear transformations are needed to form consonant clusters from AM? Ideally this set of rotations (or the more limited permutations) would not depend on our *a priori* knowledge. If we accept the more general solution of rotations, the solution is given by an eigenvalue decomposition (EVD) of the symmetrized AM $S(SNR)$ [or alternatively the singular value decomposition (SVD) of the AM]. The EVD can both characterize

---

[16] It seems that this was well understood by both Fletcher and Hartley, who both worked for AT&T.

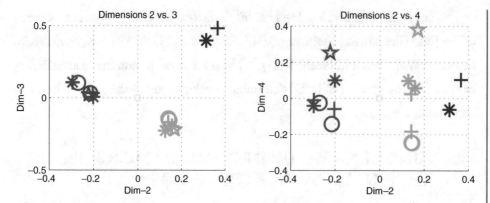

FIGURE 2.23: This figure shows a 4D symmetric approximation to the Miller–Nicely articulation matrix for the data from Table II (SNR = −12 dB). Think of the left figure as looking down on the right figure, in a 4 dimensional space. The details of this construction are described in the text.

the required rotations, *while*, at the same time, define a hierarchy of approximations to the AM, to any order of approximation. It is well known that such approximations may be generated by using the EVD expansion to reconstruct the matrix, after setting the eigenvalues to zero having magnitude less than a given threshold.[17]

In Fig. 2.23 we show such a representation for the data of Fig. 2.11 at a *SNR* of −12 dB, and in Fig. 2.24 is a similar representation for the case of −6 dB *SNR*. For these figures the CM $\mathcal{C}$ is first row normalized, and rendered symmetric, resulting in matrix $S(SNR)$, Eq. (2.27). This real 16×16 symmetric probability

---

[17] These conclusions are based on the observation that $2X = X + X^T + X - X^T$, $sort(svd(X + X^T)) = sort(abs(eig(X + X^T)))$ and $sort(svd(X - X^T)) = sort(abs(eig(X - X^T)))$, along with the fact that the matrix of eigen vectors are orthogonal. The EVD, represented by $eig(X)$, factors matrix X as $X = V\lambda V^T$ while the $svd(X)$ function factors X into $X = U\Sigma V^T$, where $\Sigma \geq 0$ and λ are diagonal and V and U are *orthogonal matrices*, defined by the property $UU^T = VV^T = 1$. Finally the U defined by $eig(X + X^T)$ and the U and V defined by $svd(X + X^T)$ are all "identical" in the following sense: $U = P_\pm V$, where $P_\pm$ is a sign permutation matrix [thus, $U = \mathcal{O}(P_\pm V)$ where $\mathcal{O}$ is *order of* within ± 1]. These sign permutations correspond to reflections across the planes perpendicular to each eigen vector. The columns of V, where $V\lambda V' \equiv X'X$, are the same as the columns of V, where $U\Sigma V' \equiv X$, within an order permutation, and sign change, when X is full rank.

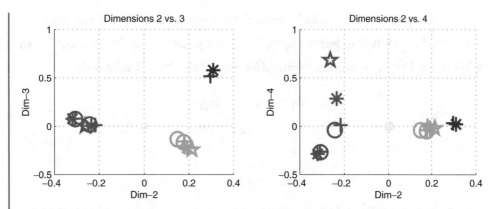

FIGURE 2.24: This figure is the same as Fig. 2.23, but for the case of −6 dB.

matrix is then decomposed into an *outer-product expansion* of the form

$$S = U\lambda U'. \tag{2.46}$$

This transformation determines a unique 16×16 real eigenvector matrix U, and a diagonal eigenvalue matrix λ, sorted in magnitude, from largest to smallest. The eigenvalue matrix λ is then absorbed into the normalization of the U eigenvectors, giving a new matrix of orthogonal but not normal eigenvectors

$$u \equiv U\sqrt{|\lambda|}. \tag{2.47}$$

An example: Let u be composed of column vectors $[u_1, u_2, \ldots, u_{16}]$. Suppose only the first four eigen values are significant. We therefore drop all but the 4 largest eigenvectors, resulting in the 16×4 matrix

$$\mathcal{U}_4 \equiv [u_1, u_2, u_3, u_4] \tag{2.48}$$

of significant rescaled eigen vectors. This allows us to approximate $S^{16 \times 16}$ to fourth order as

$$\mathcal{M}_4 = \mathcal{U}_4 \times \mathcal{U}_4'. \tag{2.49}$$

Each point in Fig. 2.23 is defined by a *row* of $\mathcal{U}_4$. Two dimensions are suppressed for each plot. For example, in the left panel we see the 16 points ψ_i in a 2D space, having coordinates defined by rows of the 16×2 submatrix

$$
\begin{array}{ccc}
\psi_1 : & u_{1,2} & u_{1,3} \\
\psi_2 : & u_{2,2} & u_{2,3} \\
\vdots & \vdots & \vdots \\
\psi_{16} : & u_{16,2} & u_{16,3}
\end{array}
\tag{2.50}
$$

defined by the second and third eigenvectors. These 16 pairs $[y_i, z_i]$ represent the rotated coordinates of the 16 sounds (in the order defined by Fig. 2.11).

Given the rotated coordinates, one may compute the Euclidean distances between each of the sounds in this 2D subspace of the transformed AM. From these distances, groups may be defined. Each group may then be characterized by its mean and variance. This method may be used when the CM is not given as a function of the SNR, as is frequently the case of highly specialized data (learning disable children, cochlear implants, etc.). A detailed description of Fig. 2.23 is presented next.

The small solid circle represents the origin. In the left panel, dimensions 2 versus 3 are shown, and we see that the confusion data cluster into three main groups. Dimension 1 is not shown since the first eigen vector is a constant, meaning that the three groups in plane 2-3 lie on a plane (i.e., have the same coordinate in dimension 1, which follows from the fact that the row-sums of $\mathcal{A}$ are 1). The left most cluster represents the unvoiced sounds. The lower cluster is the voiced sounds, while the upper-right cluster gives the two nasals. In the right panel, dimensions 2 versus 4 are displayed, and we see that the fourth dimension partially separates the sounds of each group. For example, of the UV sounds, two (/shaw/ and /saw/) are separated by dimension 4. The various symbols code place distinctions, defined as follows. ◯: [/pa/, /fa/, /ba/, /va/], +: [/ta/, /ta/, /da/, /ða/, /na/], ∗: [/ka/, /sa/, /ga/, /za/, /ma/], and ⋆: [/ʃa/, /ʒa/]. Such a representation allows us to compute the distance between each of the sounds, as a function of the dimensionality of the

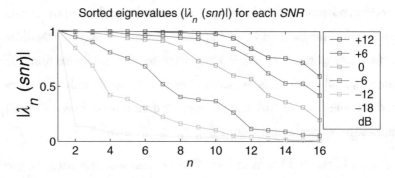

FIGURE 2.25: Eigenvalues $\lambda_n(snr)$ were computed for each Miller–Nicely AM matrix (wideband data), and were then sorted in absolute value. The resulting values $|\lambda_n(snr)|$ are shown here as a function of the eigenvalue number n with snr as a parameter.

representation. These distances are entirely based on perceptual confusion measures. The eigen values for MN55, as a function of SNR, are shown in Fig. 2.25. Arguably, these results strengthen the view that the dimensionality of the MN55 data are low, perhaps on the order of 7 to 10 dimensions (e.g., binary events).

Summary: The EVD of the symmetric part of the AM(SNR) allows us to compute the perceptual distance between each of the sounds, averaged across subjects, or the perceptual spread from the group average, across listeners, or even talkers, given $S_{s,b}$ for each case. From Fig. 2.23 one may see that the voiced, unvoiced, and nasal sounds each cluster.[18] The eigenvalue decomposition is a mathematically well defined linear family of transformations that does not depend on *a priori* knowledge. It appears to be better behaved than MDS procedures, which are known to be unstable in this application (Wang and Bilger, 1973). In fact Wang and Bilger say

> In general, the results of [multidimensional] scaling analyses support the
> notion that articulatory features, either singly, or in interactive combina-
> tion, do, in fact, constitute valid perceptual dimensions. This conclusion,

---

[18] The clusters of this figure are quite similar to the previous multidimensional scaling (MDS) clustering derived by (Soli *et al.*, 1986).

however, reflects the fact that scaling solutions invariably require considerable interpretation, and that linguistic features, as well as traditional articulatory features, provide a ready basis for that interpretation. Obtained spatial configurations have not been sufficiently stable across experiments and scaling methods to suggest definition of new and totally unanticipated perceptual features.

So far this EVD method has been less effective than working with the groupings defined by the off-diagonal elements of the CM as a function of SNR (Allen, 2005). However in special cases, the CM as a function of SNR is not available. In these cases EVD seems like the best choice of analysis.

2.7 VALIDATION OF THE AI

In this section we review and summarize the many papers that have attempted to validate the AI. Even today this empirical theory is questioned [see, for example Musch and Buus (2001a)]. *In fact, AI theory is the cornerstone of the hearing aid industry.* There is little doubt of its validity. Rather it is a question of understanding, credibility, and proper use. The most frequent error is the misapplication of AI to meaningful sounds, such as HP sentences. It is also important to remember that there are many AI procedures, the most successful, *least* analyzed, being Fletcher and Galt's (Rankovic, 1997, 2002). Others include French and Steinberg (French and Steinberg, 1947), RASTI, STI (Rapid Speech transmission Index) (Steeneken and Houtgast, 1999) and more recently the SII (S3.5-1997, 1997).

Beranek (1947): Historically Beranek's paper (Beranek, 1947) is a very important review, and was one of the first to summarize *Speech and Hearing* results following the war.[19] Near the end of Beranek's paper, comparisons are made between AI calculations and experimental data. These results show strong biases, under certain

---

[19] Prior to World War II, the two books on *Speech and Hearing* were Fletcher's 1929 seminal work and Stevens and Davis' 1938 (reprinted 1983) important overview.

conditions, with the AI scores being 10–15% higher than the data. Beranek's AI calculation was later discovered to be in error, and the problem was corrected (Pickett and Pollack, 1958; Kryter, 1962b). Beranek's analysis had failed to account for the upward spread of masking.

French and Steinberg (1947): This paper, like Beranek's, provides a broad review of the state of the knowledge of *Speech and Perception*, and then presents a method to calculate the AI. It was the first major publication to outline this theory, and was the starting point for many future studies. For a detailed history of this period see Allen (1994, 1996).

Fletcher and Galt (1952): Hundreds of combinations of conditions are presented, each of which pushes the Fletcher and Galt version of AI theory to its limits (Fletcher and Galt 1950). Many of the details are documented in several of the many Galt notebooks, which were recently obtained from the AT&T Archive, and are now available on a CDROM from the *Acoust. Soc. of Am.* (Rankovic and Allen, 2000). Many of these calculations have recently been replicated by Müsch (Müsch, 2000).

Kryter (1960–1962): One of the most widely cited papers on the validation of the ANSI version of the AI, based largely on the French and Steinberg version, is that of Kryter (Kryter, 1962b) and the companion paper (Kryter, 1962a), which defines the Kryter calculation method in fine detail, with worksheets. Unfortunately these fine papers fail to cite Fletcher and Galt's 1952 work.

Most of Kryter's results show that the 20-band method (Kryter, 1962a) worked extremely well, in agreement with French and Steinberg (Kryter, 1962b). They ran several tests: (i) narrow bands of MaxEnt speech with narrow band noise, (ii) wide band spectrally shaped noise, (iii) they recalculated data from French and Steinberg, (iv) Egan and Wiener narrow-band speech data in white and 15 dB/oct lowpass noise, along with (v) Miller's narrow-band noise maskers, and (vi) Pickett and Pollack's measurements in tilted [−12, 0, +6] dB/oct noise. Studies *(v,vi)* were important because the noise triggered the nonlinear upward spread of masking, which has the effect of attenuating signals having frequencies above those of the masker, once the masker level reaches about 65 dB SPL (Allen, 2001).

The important issue of multiple pass-band filters, for which the AI method fails, is addressed in (Kryter, 1962b, p. 1699), and even more extensively in (Kryter, 1960). These issues will be discussed further in the next section.

Boothroyd Review (1978): In 1978 Boothroyd wrote an insightful review chapter that covers some of the same topics as this review, and comes to similar conclusions. The following are summary key quotes that complement the present discussion:

> Stop consonants ... are characterized by a sudden rise of intensity after a period of silence, and the difference between nasal consonants and vowels may be perceived partly on the basis of intensity differences. (p. 120)

> The most significant finding of this research (Fletcher and Galt, 1950; French and Steinberg, 1947), and perhaps the least generally understood, [is] that under certain circumstances, the acoustic spectrum can be regarded as ... [the output of a filter bank], each [filter] contributing independently to the probability of phoneme recognition. (p. 122)

> It is clear from Miller and Nicely's work that there are marked differences in the frequency distributions of information about the different features. ... place of articulation is predominantly high frequency ... voicing and manner of articulation is available over a wide frequency range. ... these ranges correspond closely with the range of the second formant. When listening to filtered speech, however, it quickly becomes obvious that information on rhythm and intonation is, like manner and voicing information, diffusely spread across the acoustic spectrum. Much of the

information that serves to differentiate vowels from consonants and to differentiate consonants on the basis of voicing and manner of articulation is to be found in the time/intensity patterns of speech or in the relative times of occurrence of different events. (p. 127)

voiced and voiceless stops may be differentiated on the basis of the intensity of the fricative noise and the time interval between plosive release and the onset of voicing (the voice onset time). The most useful single formant is the second. (p. 127)

Duggirala *et al.* (1988):   Duggirala *et al.* (1988) used the AI model to study *distinctive features*. This is one of the few studies that looks at the effect of different frequency bands on distinctive feature error.

Sinex *et al.* (1991), looking in the cat auditory nerve, found as outlined by Abramson and Lisker (1970) that the smallest VOT-JND is coincident with the category transition (Sinex *et al.*, 1991).

Müsch (1999):   Müsch has provided a Matlab© version of Fletcher and Galt's AI calculation for free distribution (Müsch, 2000). He used this program to verify many of the conditions published by Fletcher and Galt (1950).

Stevens and Wickesberg (1999):   An interesting paper by Stevens and Wickesberg (1999) looks at the detection of voiced vs. unvoiced whispered stop consonants.

In the next sections we review several of these papers in more detail, and interpret their results, toward an improved understanding of natural robustness in HSR.

2.8   CRITICISMS OF ARTICULATION MODELS

While the AI model is heavily used today, for example in the hearing aid industry, there have been many criticisms of it in the literature. In order to fully understand the limits of this complex empirical theory, these critical comments must be addressed.

The performance factor: The performance factor *PF* was first introduced by Fletcher and Steinberg to account for long term learning effects. These learning effects are extensively documented in the internal documents from Bell Labs (Rankovic and Allen, 2000). There has since been much uncertainty about the significance of the *PF*, and it has been seen by some as a free parameter that can be used to fit almost any monotonic curve. This is far from the case, and is an unfair representation. As shown by Allen (1994, 1996), *PF* is constrained by either the maximum articulation (or minimum error), under the very best of conditions. The parameter $e_{min} = 1 - s_{max}$ [Eq. (2.10) with $AI = 1$] is equivalent to the PF. It is essential to measure this maximum articulation point for new materials. The maximum articulation may be largely dominated by production errors (unpublished observations of the author) (Kryter, 1962b).

Hirsh (1954):   Hirsh distinguishes the testing of communication equipment (for normal hearing) from clinical measurements (on the hearing impaired). A major thrust of this study was in relating the Intelligibility to the Articulation scores. His main criticism was that for a similar AI value, filtering and noise give different scores, with higher scores for the noise than for filtering.

The paper found relations in the data (i.e., Hirsh's Fig. 10) that are similar to those found by Boothroyd and Nittrouer, and which may be modeled by Boothroyd's *k*-factor. Specifically, as the noise and filtering is increased, $W(S)$ of Fig. 3.1 applies, with a *k*-factor that depends on case. For filtering $k \approx 7$, while for noise $k \approx 3$.

There are several things we can say about this analysis. First the AI was not intended to be used with meaningful words, and the more we understand about context effects, the more clearly this rule applies. From the work of Miller's, Boothroyd *et al.*'s, and Bronkhorst *et al.* on context, it is clear that the AI is a measure of the front end event extraction process. Thus the measure given in Hirsh's Fig. 10 is *not* a test of the AI because of the influences of context on his experimental results.

Rankovic tested the question of filtering and noise explicitly (Rankovic, 1997) and found that the ANSI AI standard treated filtering and noise differently, as found by Hirsh. When she used the full Fletcher calculation on her own database of MaxEnt sound articulations, the agreement was nearly perfect. Thus she confirmed[20] that the Fletcher and Galt (1950) AI deals with this problem, for the case of MaxEnt syllables. Hirsh *et al.* raise an important issue, but maybe for the wrong reasons, since they did not use MaxEnt sounds.

Kryter (1962):   As pointed out in Section 2.7, Kryter found conditions that did not agree with the AI, as discussed in several papers (Kryter, 1960, 1962a, 1962b). Kryter saw a 15% higher measured score than that predicted by his AI computation (Kryter, 1962b) when the midband was removed. Kryter's results may be summarized as follows. The two band errors e_1, e_2 and the wide band error e are measured. The AI would predict that the wideband error should be (ignoring the masking of the low band on the high band) $e_1 e_2$. However, Kryter found a total error that was smaller than this (i.e., $e < e_1 e_2$), and an AI that under-predicted the measured articulation. The source of this large 10–15% articulation underestimate for the AI has remained elusive for these multi-bandpass and stop-band conditions.

One possibility is that the information in the bands next to the dead-band are not properly extracted, due to the fact that neighboring channels need to be present for processing to function normally. By removing the band, the information in the edges is lost. The AI continues to include this information, but the brain fails to use it. This would cause the AI to over-predict the actual score. This conjecture is not supported by any specific data that I know of.

---

[20] Rankovic found an AT&T internal memo, written by Fletcher (July 6, 1949-1100-HF-JO), discussing the Fletcher and Galt paper, which clearly states that the French and Steinberg method is not adequate with a system "having intense pure tone interference or having certain kinds of distributed noise interference as exemplified by some of the Harvard systems." The letter goes on to state that the Fletcher and Galt calculation has solved this problem.

Warren *et al.* (1995–2000):   Warren and colleagues did a series of studies of speech intelligibility of CID HP sentences in very narrow bands. In their 1995 study, a bandpass filter was created by cascading two highpass and two lowpass 48 dB/oct electronic filters, centered on 1.5 kHz, forming a 1/3 octave band.[21] They found that the speech key words were 95% intelligible (Warren *et al.*, 1995). This level of intelligibility is in conflict with all other studies, and with computations of the AI, which give a much smaller articulation. Fletcher and Galt found 12.5% and Egan and Wiener 25% for MaxEnt CVCs.[22] Warren *et al.* also looked at pairs of bands, and concluded with the following claim:

> The increase in intelligibility for dual-band presentation appears to be more than additive.

This statement is strange since the whole basis of the AI is to find transformed articulations that are additive. The intelligibility is known *not* to be additive (Fletcher, 1921, 1995). Warren *et al.* also concluded, without actually computing the AI, that

> AI measurements cannot be used to predict the intelligibility of stand-alone bands of speech.

Clearly something is wrong somewhere. First, they should have computed the AI for their conditions. Second, they did not add low level masking noise. Third they used HP sentences.

Fortunately in follow up studies the problem has been resolved. There were two problems with the 1995 study: First was the use of HP sentences, and second was the amount of speech information in the skirts of the filter. Once the

---

[21] The actual bandwidth is not given in the *methods section*, however several of the follow-up studies state that the bandwidth was 1/3 oct. A 1/3 octave filter centered on 1.5 kHz has a "3-dB down" bandwidth of 345 Hz.

[22] Fletcher and Galt report an articulation of 25% for MaxEnt CV phones (the bandwidth was ≈800 Hz) (Fletcher, 1995, H-8, Fig. 212, p. 390). Thus $s \approx \sqrt{0.25} = 0.5$, requiring that CVCs would be 1/8, or 12.5% correct (0.5^3). Egan and Wiener (1946) found MaxEnt CVC syllables to be 25% correct in a bandwidth of ≈630 Hz (Egan and Wiener, 1946, Fig. 2, p. 437).

HP sentences were replaced with MaxEnt words, the CVC score dropped to the value previously observed by Egan and Wiener (25%). Namely Bashford *et al.* found a score of 26% for the 1/3 octave 96 dB/oct filter. After the electronic filters had been replaced with very high order digital filters, the score dropped to 4% (Bashford *et al.*, 2000).

Thus when executed properly, there is no disagreement with the AI, and no "more than additive" effect. Unfortunately there were many years of uncertainty introduced by the 1995 results, which probably should not have been published without first doing the obvious sanity checks. When a result is out of line with the rest of the literature, the burden of proof is on the shoulders of the presenters of the unexpected results.

Lippmann (1996):   Lippman measured articulation scores for MaxEnt speech filtered into two bands: a fixed lowpass at 800 Hz along with a highpass at a variable frequency (f_c). The variable highpass channel was one of [3.15, 4, 5, 6.3, 8, 10, ∞] kHz. In Fig. 2.26 we use 12 kHz as a surrogate for $f_c = \infty$, corresponding to the high band turned off.

Lippmann's claim is that this stop-band data does not agree with the Articulation Index, yet he did not compute the AI. To test Lippmann's claim, the AI was computed for the conditions of the experiment, as shown in the Fig. 2.26. Table 63 on p. 333 of Fletcher 1953 book was used. The values of the table were interpolated to get $A(f_c)$ at high resolution, and then added in the proportion indicated by Lippmann's filter cutoffs. When computing the two band edges, a factor of 1.2 times the 800 Hz lowpass frequency, and the highpass frequency divided by 1.2 were used as the relevant cutoff frequencies. This is similar to the technique previously used (Kryter, 1960). The formula for $P_c(A)$ for this calculation was Eq. (2.10). As may be seen from Fig. 2.26, the agreement is quite good for the lower cutoff frequencies, up to 8 kHz, but is not good for the low band alone (labeled, for convenience, as 12 kHz), where the AI gives 65% and the Lippmann data is 44.3%. Fletcher reports an experiment score of 70% (Fletcher, 1953, Fig. 199, p. 384) for

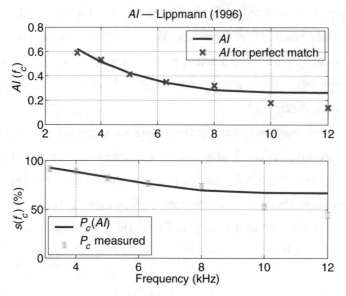

FIGURE 2.26: Lippmann (1996) measured the consonant error for MaxEnt sounds with two bands, similar to Kryter (1960). The first band was a lowpass at $f_c = 800$ Hz. The second band was for a high filter at a frequency given by the abscissa Lippmann (1996).

MaxEnt phonemes lowpass filtered to 800 Hz (1 kHz at the 40 dB down point on the filter skirt). Lippmann's scores go *below* the Fletcher and Galt observations when the high band is added at 10 kHz, and when no high band is provided at all.

The AI was designed to predict the average CVC score $s = (2c + v)/3$. Presumably Lippmann reported scores are for the average of the initial and final consonants ($s = (c_i + c_f)/2$). Vowel scores are not reported. Normally adding the vowels would make the score higher, because the vowels are more intense and therefore have a much higher *SNR*. The vowels depend on the mid-band that was removed for this experiment. If more carefully analyzed, this experiment might give some insight into the frequency regions corresponding to the different vowel sounds.

The results of this short paper are tantalizing in that they suggest possible insights into phone recognition. For example, what are the scores for the vowels with the mid band missing? How do the scores for the initial and final consonants differ? What is v/c for these data. Fletcher's analysis (see his discussion of "factor X" on pp. 286 and 298 of his 1953 book) showed the importance of this ratio when defining the average phone statistic $s = (s_i + v + s_f)/3$, which may not be a good statistic for Lippmann's experiment.

Müsch and Buus (2001):   This paper is an interesting bold attempt to account for the 1951 results of Miller, Heise, and Lichten, based on the earlier modeling work of Green and Swets, supplemented by the multichannel signal detection model developed by Durlach *et al.* (Green and Swets, 1988; Durlach *et al.*, 1986). The approach is based on a detection theory model of "one of M–orthogonal Gaussian signals" (MOGS) (Green and Swets, 1988, p. 307). Signals are said to be "orthogonal" if, under the best of conditions (i.e., as $d' \to \infty$), the M×M detection task AM is diagonal.[23] Hartmann used identical mathematics to model (M + 1)-interval experiments (Hartmann, 1997). This theory (MOGS) calculates the average $P_c(d')$ for the detection of M independent Gaussian distributed waveforms. The resulting psychometric function [i.e., $P_c(d')$] is 2^{-M} when $d' = 0$, and goes to 1 as d' becomes large. Since this is the basic character of the Miller *et al.* data, as shown in Figs. 2.6 and 2.7, Müsch and Buus speculate [following those of (Green and Swets, 1988)] that the MOGS model might apply, or have useful application, in providing insight into the Miller *et al.* (1951) data.

  The Müsch and Buus article is directed at highlighting several problems with AI. On page 2897 they state that AI was developed to predict material with context:

> Speech-test scores expressed as percentage of correct responses are derived from the AI through a transformation that is test-material and test-format specific. Different transformations must be used when the number of items in the set of test stimuli changes. Different transformations must also be used for different levels of constraints placed upon the

---

[23] The quantity d' is the ratio of the difference in means of two distributions (i.e., typically Gaussian distributions) divided by the standard deviation. It is the same as a statistical "t-test," in its most elementary form.

message set. For example, low-predictability sentences require a different transformation than high-predictability sentences. Consequently, a
large set of transformation functions is needed to accommodate the large
number of possible speech-material/test-format combinations. . . . With
the exception of a transformation for open-set phoneme recognition in a
nonsense context (Fletcher and Galt, 1950), all of these transformations
are derived empirically.

This statement seems in conflict with the development of the AI model, which
was developed for open-set MaxEnt ZP sounds (the Fletcher and Galt exception
noted in the quote). It was specifically not designed to deal with HP sentences.
While the ANSI standard provides $P_c(A, M)$ functions for several vocabulary sizes
M, to account for context effects, the severe limits of AI when used on high context words and sentences is carefully described in Kryter's papers (Kryter, 1962b,
p. 1697). In fact Boothroyd develop his k- and j-factor methods to specifically
account for such context effects, starting from an AI calculation or experimental
measure (Boothroyd and Nittrouer, 1988). If the AI had been able to deal with HP
sentences, then Boothroyd's papers and method would be unnecessary.

One must conclude that the MOGS model is accounting for the poorly
understood affects of chance and context in the Miller, Heise and Lichten data,
rather than problems with the AI method (Eq. (2.12)), as stated by Müsh and Buus.

The paper introduces the concept of *synergistic and redundant* interactions,
based on the work of Warren, Kryter, Hirsh *et al.*, Lippmann, and others.

one such shortcoming is the AI model's inability to account for *synergistic
and redundant interactions* among the various spectral regions of the
speech spectrum. [p. 2896; emphasis added]

The *synergistic and redundant interactions* they speak of stem from the criticisms of
the AI discussed earlier in this article (Warren, Kryter, Hirsh *et al.*, Lippmann,
etc.). Their term *synergistic* is defined in their sentence:

> When the error probability in a broadband condition is less than the product of the *error probabilities* in the individual sub-bands, the bands interact with *synergy*. [p. 2897; emphasis added]

while the term *redundant* is defined as

> when the error probability in the broadband condition is larger than the product of the error probabilities in the individual bands, the bands are considered to carry *redundant information*. [p. 2897; emphasis added]

These two definitions are presented graphically in Fig. 2.27.

For the *synergy* case ($e < e_1e_2$), the measured error $e(SNR)$ is smaller than the model error ($\hat{e} = e_1e_2$). This is what one could expect if there are bands containing independent information (active channels) missing from the calculation, since if these missing bands were factored into the prediction, the discrepancy in error would decrease. As an example, if there were an visual side channel, the true error would necessarily be smaller than what the auditory model would predict. Thus *synergy* may be thought of as a deficient model, having missing channels.

For the second definition, which Müsh and Buus define as *redundant informa-tion*, $= e_1e_2$, the model error $\hat{e} = e_1e_2$ is smaller than the measured error $e(SNR)$, implying that the band errors themselves are smaller than they should be. This would be a glaring error, since it would mean either that the observer is either not integrating the information across bands (when they listened to individual band, they heard the cue, but when they listened to all the bands, they missed it, making the wideband error greater), or the bands have overlapping *common* information, that cannot be used to advantage, because the observer already got the information from another band. This would truly be "redundancy" across bands.

The real question is, do these "synergistic and redundant interactions" actually exist in the data? This is a challenging and important question. Müsch and Buus base their case on the earlier work of Hirsh, Kryter, Warren, and Lippmann. This

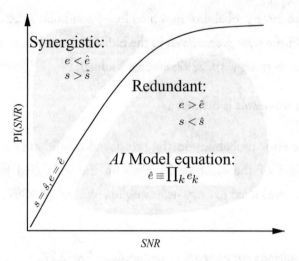

FIGURE 2.27: When the measured phone score $s(SNR)$ is greater than the AI model prediction $\hat{s}(SNR)$, Müsch and Buus call the model *Synergistic*, and when the model predicts a score that is less than the measured score, it is *Redundant*. This scheme is conceptually a nice way of dealing with error in the AI model, but we must be careful not to lead the interpretation of such model deviations. For example, in the case of the Kryter-effect, a dead-band results in $s < \hat{s}$, and therefore leads to "redundancy." This sounds a bit like the Kryter-effect is explained, when it is not. How could the introduction of a dead-band somehow create redundancy in the data? A data-centric terminology might be clearer. For example, one might simply say that the model over- or under-predicts the score. For the Kryter-effect case "The AI model over-predicts the score."

review has discussed each of these papers, and challenged identifiable deviations from AI theory for Hirsh, Warren, and Lippmann. Kryter may have identified an interesting redundant effect, which is still in need of an explanation, for his case of 2–3 noncontiguous bands of speech, with no noise present. However until these conditions have be tested with the Fletcher and Galt model, we cannot be sure. In any case, the use of MOGS to fix AI synergy or redundant effects due to context [i.e., the results of (Miller *et al.*, 1951)], seem problematic, since AI does not the "synergistic" effect of context (Boothroyd and Nittrouer, 1988).

C H A P T E R   3

Intelligibility

Other than for the brief digression in Section 1.3.1 regarding Fig. 1.4, we have concentrated on articulation studies which take a bottom-up view. We next turn to studies investigating top-down word recognition, using meaningful words and context.

The recognition chain: Figs. 1.3 and 3.1 summarize the statistical models that take us from the speech signal, to band SNRs, event, phone, syllable, word, and sentence probabilities. Once the band SNRs are determined, the event errors e_k are computed, then the mean phone score s, and then the syllable articulation S. Given S, the meaningful word probabilities W, $\mathcal{I}$ (ZP sentence scores), and $\mathcal{W}$ (words in HP sentence) may be computed. Each of these models has one free parameter, such as $e_{\min}$, k_1, k_2, and j_1.

In Fig. 1.3 we associated the band errors with elementary events obtained from the parallel processing of the cochlea and, say, the auditory cortex. We chose the auditory cortex here without much evidence other than the fact that it is known that tonotopic maps are present in this cortex, so phones could not have been recognized up to, and including, these layers of processing.[1] The band errors may represent probabilities of the detection of binary events extracted from the speech signal. If any (or many) of the channels capture the event, the feature is recognized

---

[1] For our purpose here, we do not need to know where the event is extracted.

- The cochlear critical bandwidth defines the internal SNR_k
- The *event-error* model: $e_k \propto e_{\min}{}^{AI_k}$ (AI is in dB units)
- The *average-phone articulation* model: $s = 1 - e_1 e_2 \cdots e_k \cdots e_K$
- The MaxEnt CV *syllable articulation* model: $S = s^2$
- The MaxEnt CVC *syllable articulation* model: $S = s^3$
- Heuristic degree of freedom *context models* Boothroyd & Nittrouer (1988)

  Words from MaxEnt syllables: $W(S) = 1 - (1-S)^{k_1}$

  ZP sentences: $\mathcal{I}(W) = W^{j_1}$

  Words in HP sentence: $\mathcal{W}(\mathcal{I}) = 1 - (1-W)^{k_2}$

  HP sentence (having context): $\mathcal{C}(\mathcal{I}) = 1 - (1-\mathcal{I})^{k_3}$

- Layers of context

  k_1 depends on the number of meaningful words in MaxEnt words.

  j_1 depends on the word salience for the topic context,

  k_2 depends on the context effects due to semantics and grammar

FIGURE 3.1: The recognition chain. This set of statistical relations summarizes the relations shown in the lower part of Fig. 1.3. The probability correct for words is defined as W. The probability for recognition of words in high predictability sentences (HP) is defined as $\mathcal{W}$. Thus the measure $\mathcal{W}(W)$ results from the HP sentence context. An example of a ZP sentence is "Sing his get throw." An example HP sentence is "Warm sun feels good."

with a probability that depends on the *SNRs* in the many channels. When sufficient features are determined, the phone is identified, as measured by probability s.

In summary, context effects seem to boil down to complex problems of combinatorics, based on the probabilities of elementary (sub-phonemic) events. This is an important conclusion that, when properly understood, can significantly impact our view of ASR. One important implication of the model (i.e., Fig. 1.3) is that, once a few model parameters have been established, the band SNRs are all that are required to establish the HP sentence error score.

3.1   BOOTHROYD (1968–2002)

The results of Boothroyd extend Fletcher and Stewart's "independence model," and thus the AI method, by looking at the effects of context, both at the word level and the sentence level (Boothroyd, 1968, 1978; Boothroyd and Nittrouer, 1988). There are many fewer CVC words than there are CVC sounds, creating a context effect. In his 1968 paper, Boothroyd states that "the effect of context is quantitatively equivalent to adding statistically independent channels of sensory data to those already available from the speech units themselves." This assumption leads to the parallel processing relation (denoted the "k-factor")

$$W = 1 - (1 - S)^k, \tag{3.1}$$

where $k \geq 1$ is a degrees of freedom parameter, S is the MaxEnt syllable articulation score, and W is the meaningful word recognition score. This equation takes the score of elements (i.e., S) measured under conditions of no context, and returns the score of the same elements (i.e., W) after context has been added. A value of $k = 1$ corresponds to no context, giving $W = S$.

Boothroyd suggests we view Eq. (3.1) as a the product of two correlated error terms, by factoring the error as

$$W = 1 - (1 - S)(1 - S)^{k-1}. \tag{3.2}$$

When written this way the context may be viewed as an independent context channel having probability $(1 - S)^{k-1}$. The context information goes to zero as $S \to 0$. Thus the context may be viewed as an independent parallel processing channel.

A second formula is used to predict "from *elements* to *wholes*." When word scores W (an element) is use to predict a sentence score $\mathcal{I}$ (whole), Boothroyd found a formula similar to sequential processing relation Eq. (2.3), but with an exponent that depends on the number of functional words (i.e., the degree of the context effect), which he called a "j-factor"

$$\mathcal{I} = W^j. \tag{3.3}$$

The j factor is between one and the number of words in the sentence. For MaxEnt sentences [what Boothroyd calls "zero predictability" (ZP) sentences], j equals the number of words. For "high predictability" (HP) sentences, j is less than the number of words. (Example with $j = 1$: Do you like sweets? I do *not* like them.)

The 1988 paper gives several plots showing the individual points on the curve, which give an idea of the magnitude of the error for these models. The authors also show the effects of cascading the various models to predict meaningful (HP) sentences from MaxEnt phones and words.

Grant and Seitz have recently estimated k values (parallel processing) under various conditions of sentence context (Grant and Seitz, 2000), with k in the range from 1.5 to 2.25.

Recently it has been reported that Boothroyd's "j" factor may only be valid under circumstances where the listener is under listening stress. This is a most interesting possibility that could have an important impact on any future modeling of language context effects (Chisolm, personal communication, 2000; Boothroyd, 2002; Grant and Seitz, 2001; Bronkhorst *et al.*, 2002).

3.2   BRONKHORST *ET AL.* (1993)

This important paper takes a forward looking approach to speech context effects, developing a specific and detailed probabilistic model of context processing. The model is then verified with an acoustic and an orthographic experiment, giving a deep insight into how human context processing works. An important application includes a simulation of the Miller, Heise, and Lichten experiment (Miller *et al.*, 1951). At the heart of this research is a generalization of the j and k context models of Boothroyd (sequential and parallel processing), which also accounts for chance.

The formulation assumes a *two-stage* recognition process: *First* the elements are recognized, and second context is used to fill in missing elements. The first stage represents input, while the second stage contains the actual model. This two-stage approach is consistent with the outline of HSR described in Fig. 1.3. The front-end input is either derived from the AI, measured, or designed into the experiment, as in the case of the orthographic experiment. Unlike Fig. 1.3, the model is specific. In the orthographic experiment, the results are explicitly modeled, with the outcome that human and machine performance are functionally identical. This is a proof, by example, of the claim that context processing may be done by a deterministic state machine. No model of the first stage (the auditory front-end) is attempted.

An example: A simple example of the Bronkhorst *et al.* model (the second stage) for the special case of a CV is instructive, and will suffice for the present discussion. There are two sounds in a CV, and the articulation for the consonant is c, and for the vowel is v. There are four possible outcomes from a trial, both correct, C correct V incorrect, V correct C incorrect, and both incorrect. In the model the outcomes are grouped into the *number of* errors (n) that are made in the syllable, giving three measures, for our CV example, called Q_n with $n = 0, 1, 2$:

$$Q_0 = cv$$
$$Q_1 = (1 - c)v + c(1 - v) \qquad (3.4)$$
$$Q_2 = (1 - c)(1 - v).$$

Assuming *independence*, the articulations are multiplied. The first term Q_0 is the probability of 0 errors, Q_1 is the probability of making 1 error, in any position, while Q_2 is the probability of making 2 errors. These measures are the outputs of the first stage of processing (i.e., they represent inputs to the model), and correspond to the models used in the formulation of Fletcher's AI, such as Eq. (2.4). The term Q_0 corresponds to sequential processing, while Q_2 corresponds to parallel processing. The term Q_1 is novel, and has not been previously considered in any context model. These equations summarize the first (front-end) stage of processing. Either c and v or the three Q_n values are the inputs to the second stage (the model).

Context is accounted for by the model (stage 2). This is done by introducing probabilities c_1, which represent the chance of guessing either C or V, and c_1c_2, which represents the chance of guessing the CV, when neither the C or the V is heard. Probability $c_1 c_2$ can be taken to be $1/N_{cv}$, where N_{cv} is the cardinality of the set of CV sounds, or zero, depending on the details of the experimental measurement. The output of the model is the context model for the CV recognition

$$W_{cv} = Q_0 + c_1 Q_1 + c_1 c_2 Q_2. \tag{3.5}$$

The variable W_{cv} indicates the intelligibility after context has been accounted for (namely the intelligibility at the output of the model, while $\{c, v\}$ is the articulation at the input to the model), and explicitly models the last two boxes of Fig. 1.3. For the case where $c_2 = 0$ (chance of guessing the CV word is zero), the context model equation would be

$$W_{cv} = Q_0 + c_1 Q_1. \tag{3.6}$$

A model equation is also given for recognizing one element (i.e., c or v, which we shall call s), which in this case is

$$s = (1 - c_1) Q_1 + 2(1 - c_1) c_2 Q_2. \tag{3.7}$$

Again, when $c_2 = 0$, and writing this expression in terms of $q's$ gives

$$s = (1 - c_1)(1 - c)v + (1 - c_1)c(1 - v). \tag{3.8}$$

The terms $(1 - c_1)(1 - c)$ and $(1 - c_1)(1 - v)$ represent the probability of getting one unit wrong (C or V) and then guessing that unit, all under the assumption of independence. These formulas are related to Eq. (2.2), and allow us to explore $\lambda(SNR) = v/c$. Figure 2 of the Bronkhorst $et\ al.$ paper finds that

$$v = 1 - (1 - c)^k, \tag{3.9}$$

with $2.6 \leq k \leq 3.9$. Solving for λ gives $\lambda \approx \sqrt{k}$.

Finding the c_i's: The inputs to the Bronkhorst $et\ al.$ model are the c and v, which are used to compute the Q_n's. The other required input parameters, the c_i's, which are directly computed from a dictionary. The values of c_1, c_2, and c_3, based on a CVC dictionary, are shown in Fig. 3 of the Bronkhorst *et al.* paper.

Coarticulation effects: A most interesting problem this paper tacitly raises is the question of the aural information bearing element: is it in the consonant and vowel, as basic units, or is it in the transition (the CV and VC transition)? For example, given a CV, there is potentially just as much information in the C,V combination as there is in the CV transition. Suppose their are 2 C and 2 V sounds. Then the total number of possibilities are $2 \times 2 = 4$ CV sounds. The number of transitions is 2 to 2, which is also 4. Thus if we code the information as a transition, or more directly as C and V units, the total entropy could be the same. This is not a minor point. We propose the following definitions: We call a 2-phone the grouping of C followed by V (or V followed by C). We define a *diphone* as the CV unit, namely where the information is carried by the *transition* between the initial and final consonant. The issue here is about the relative entropy in diphone vs. 2-phoneme coding. An orthographic representation is a unit code. Speech could be a diphone code, and yet be represented orthographically as a unit (phoneme) code, because, in theory, the entropy could be the same. As evidence for diphone coding, single letter diphthong vowels, such as "I" and "a," are VV sounds, not stationary vowels.

3.3   TRUNCATION EXPERIMENTS AND COARTICULATION, FURUI (1986)

In 1986, Furui did a series of experiments, which gives insight into the temporal processing of events and perhaps coarticulation. It seems that the CV events are very compact in time, maybe in as little as 10 ms. In Furui's experiments, CV sounds were truncated from the onset and from the end, and listeners were asked to identify the truncated sound. In Fig. 3.2 some of the identification results are shown. In the left panel we see the C, V, and CV scores as a function of truncation from the onset, and in the right panel we see the C, V, and CV scores when truncating from the end of the sound.

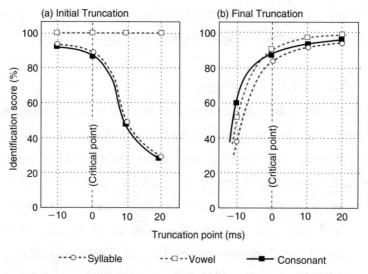

FIGURE 3.2: In the left panel of this figure from Furui (1986) the CV sounds is truncated from the onset. At some time, labeled the *critical point*, the consonant identification score drops from 90% to 45% within 10 ms. The vowel score is constant at close to 100% for these truncation times. In the right panel the CV is truncated from the end. In this case both the C and V change with nearly the same "critical point." In one example shown in the paper, the two critical point are within 10 ms of each other. The most obvious interpretation of this data is that the CV sounds carries the information in a well-localized time having ≈10-ms duration.

3.4   VAN PETTEN *et al.* (1999)

This study (Van Petten et al., 1999) [see also (Van Petten and Kutas, 1991)] sets out to find how long it takes to process semantic context. The first step in this study was to determined the *isolation point* (IP) for 704 words in neutral (ZP) sentences. The IP is defined as the time from the start of the word, quantized to 50 ms steps, such that 7 out of 10 people recognize the word. Each of the 704 words was placed in the neutral carrier sentence "The next word is *test-word*." The truncated words were randomized with all the other words, and then presented to the listeners. In the left panel of Fig. 3.3 we see the average of all the scores, synchronized to the IP determined for each word. In the right panel we see the distribution of IPs and

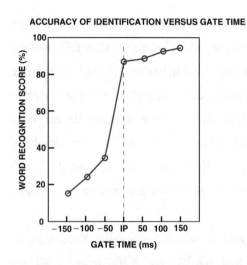

ACCURACY OF IDENTIFICATION VERSUS GATE TIME

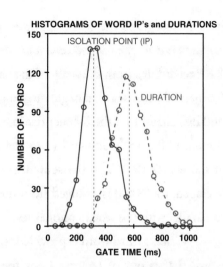

HISTOGRAMS OF WORD IP's and DURATIONS

FIGURE 3.3: The *isolation point* is defined as that truncation time corresponding to more than 70% correct recognition, quantized to 50-ms intervals. The corpus of 704 words were measured in this experiment. This figure shows the average over all 704 words, synchronized to each word's IP.

word durations. The mode of the IP is close to 300 ms, while the mode of word of durations is just under 600 ms. It seems that words are typically recognized half way through the word, well before the talker has finished the utterance.

When the words are placed in a simple context, as defined in Fig. 3.4, the listeners identified that context with the same timing and temporal resolution as

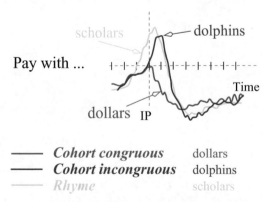

FIGURE 3.4: Averaged ERP N400 responses synchronized to the IP, for many rhyming word responses. The time graduals are 200 ms apart in this figure.

used to recognize the word. In other words, *the recognition of the words and the context must be done in parallel, and simultaneously.* This conclusion follows from ERP recordings from the scalp of the listener. A brain wave signal called the N400 (so-named because it is a negative going scalp potential, with an average latency of 400 ms) shows that the listener distinguishes the difference between the *congruous* sentence "Pay with dollars." from the *incongruous* sentence "Pay with dolphins" at the word's IP. When the sentences are synchronized to the IP and the ERP's averaged, the N400 is resolved from the noise. It follows that context is processed in parallel with the word's phonemes.

When a third word is provided (scholars), which rhymes with the congruous word, that is out of context from the first syllable, the N400 triggers 200 ms *before* the IP. These results are consistent with the idea that context processing is being done in parallel with the phone and word recognition. The point most relevant to the present paper is that the timing of the use of context information is quite restricted, and is done in "real time," with no significant delay (i.e., somewhere between 50 and 200 ms).

CHAPTER 4

Discussion with Historical Context

There is a lot to be learned by reviewing the speech articulation literature. Starting from the invention of the telephone, speech testing became a critical exercise. This research began with Rayleigh in 1908 and blossomed into the work of George Miller in the 1950s. Computers have done a lot for speech testing, but they may have provided a distraction from the important speech testing. With the computer revolution now maturing, we can return to the real issues of human information processing, and the many important questions regarding the articulation matrix (AM). Of course, today's computers make otherwise difficult speech testing almost trivial.

Following the introduction, some fundamental definitions are introduced that relate speech perception, communication theory, and information theory. A key concept is the AM, composed of all possible PI functions $P_{s,h}(SNR)$, which are defined as the probability of correctly hearing sound h after speaking sound s, as a function of the SNR. A basic model of HSR is introduced in Fig. 1.3 shows the cochlea followed by a cascade of processing layers. The initial layers represent analog across-frequency processing, defining the *front-end*. The front-end determines the underlying perceptual speech features, denoted *events*, which are the source of the quantal recognition blocks of the AM at intermediate signal to noise ratios. Quantal effects in the AM were first seen by Campbell (1910), but were first systematically studied by Miller and Nicely, as shown in Fig. 2.12, based on AM data similar to the data of Fig. 2.11. When such AM data are analyzed by eigenvalue decomposition, we see a natural clustering of the sounds, as shown in Fig. 2.23, reflecting the underlying events.

A much more precise method of finding the groups has been developed by Allen (2005), and further developed in this review. By finding those sounds h for a given s, for which $P_{s,h}$ *(SNR)* rises above chance as the *SNR* increases from its lowest levels, it is possible to precisely define every group. (Those sounds h which do not have rising PI functions are not in that group.) The relative strength of each group may be measured by the corresponding local maximum in the off-diagonal PI functions, as seen in Fig. 2.14, or in AI_g when plotted as a function of the AI, as explicitly shown in Fig. 2.19. These methods have resulted in a reliable way of defining the AM groups.

The model provides us with a way of understanding and quantifying robustness in speech recognition. Many years of study and modeling have clearly demonstrated that existing ASR systems are not robust to noise, factors due to the acoustic environment (i.e, reverberation and channel frequency response), and talker variations (respiratory illness, mouth to microphone distance, regional speaker dialects, cultural accents, vocal tract length, gender, age, pitch). Robustness to these factors is the key issue when comparing ASR and HSR performance. The question raised here is "*Why* is ASR so fragile?", or asked another way, "Why is HSR so robust?".

The HSR AM holds the answer to the robustness question. An analysis of the classic data of Miller and Nicely (1955) provides insight into the details of what is going on in HSR. This analysis is incomplete however. Much more research will be necessary before we can emulate human robustness. Quantifying it is a necessary first step.

This still leaves us with the important and difficult job of identifying the underlying events that give rise to each group. One important goal of this monograph is to establish with absolutely clarity that the next research frontier is to determine the events. More research must be done to find ways to do this, and this work is in progress.

Once the *events* have been determined, we will be able to move into the domain of discrete processing, the *back-end*, modeled as a *noiseless state machine*. This term implies that the discrete input events produce a deterministic discrete output. According to the HSR model, all the errors are due to front-end event processing. An important interface is the point where the results first become discrete, as represented by the "???" in Fig. 1.3.

No review of speech intelligibility and articulation can be complete without a thorough discussion of Fletcher's AI, reviewed in Section 2.1. The AI is direct descendent of Rayleigh's and Campbell's experiments. These experiments, which were driven by the theory of probability, are also a natural precursor to the development of *information theory*. Starting from an AT&T internal publication in 1921, Fletcher developed two classes of articulation relations. The first class was products of probability correct, as in the relations between the mean phone score s and MaxEnt syllables S, as described by Eqs. (2.2)–(2.4). I have called this class *sequential processing*, to reflect the fact that any error in a chain reduces the score. The second class is what I have called *parallel processing*, corresponding to across-frequency listening, as given by Eq. (2.12). The formulation of this relationship resulted from an assumption of *additivity of articulation* Eq. (2.6). In parallel processing, a lone channel, of many, can increase the articulation to 100%. Channels having no information do not contribute to the final results. It is convenient to think of all possible outcomes as a unit volume, as depicted in Fig. 2.5. Such a geometrical construction helps us to think about all the possible permutations.

Steinberg, who had worked with Fletcher on AI theory for almost 20 years, and French published their classic paper in 1947, after being challenged by Leo Berenak to make their method public. In this paper they provided the formula for the band errors e_k given by Eq. (2.14). Section 2.5.5 has shown, for the first time, that the AI, defined by Eq. (2.42), is closely related to Shannon's formula for the channel capacity for a Gaussian channel [Eq. (2.43)].[1]

---

[1] It is fascinating that the early formulation of information theory, by Hartley and Nyquist, and Fletcher's AI theory, were being simultaneously developed at the same company. It is proving

After the publication in 1938 of S. S. Stevens' book on basic psychoacoustics, loudness and intensity discrimination became a main topic of research at Harvard. Soon, because of World War II, AI and speech testing became the focus. Much of this work, at Harvard and MIT, was soon the charge of George Miller, who clearly was intellectually in command of the topic. He wrote many of his classic papers and books during and following this war effort. Prior to 1950, almost all the speech testing was open-set testing. Miller introduced the closed-set test, in an attempt to tease apart the role of *Entropy* in the articulation task.

Miller frequently cited Claude Shannon and his theory of information. Shannon had crystallized many of the ideas that had been floating around Bell Labs, such as Nyquist's sampling theorem for band limited signals, and Hartley's ideas on counting the number of levels in terms of the number of standard deviations over the range of the signal (Hartley, 1928). This idea of counting intensity JNDs was well known to Hartley (a Rhodes Scholar) from his knowledge of classical psychoacoustic principles (Hartley wrote one of the first modern papers on localization (Hartley and Fry, 1921)).

Miller introduced these basic tools, information theory, entropy, and the discrete channel, into his speech perception studies. Once the AM was measured in sufficient detail, it was clear that something more basic than a phone (or phoneme) were present in the data. Miller and Nicely's results confirmed and supported the observations of Halle (Jakobson *et al.,* 1967), as well as work at Haskins Laboratory on spectrogram speech synthesis, that some sort of basic *distinctive features* were being used as an underlying speech code.

In my view, Miller and Nicely's insights were sidetracked by two research misdirections. *First* was the distinctive feature, and *second* was the research

---

difficult to make the connection however. Hartley was on medical leave from 1929–1939. Nyquist was in AT&T's Development and Research (D&R) organization and, after it was merged with Bell Labs in 1934, went to *Transmission development* (area 34), Department 348. Fletcher was in Western Electric Engineering until Bell Labs was created in 1929. In 1930 Fletcher was promoted to *Director of Physical Research* (328) in the research organization (area 32), within the newly formed Bell Labs. This promotion reflected his many important contributions.

on language modeling. While both of these areas of research have their important payoffs, neither was directed at the larger problem of ASR robustness. It is the lack of robustness that keeps ASR from moving out of the laboratory. In my view, the key research question is "How does the auditory system extract perceptual features (i.e., *events*)?". By using the tools of linear algebra, such as eigen and singular value decomposition, and the off-diagonal PI function analysis described in Allen (2005), one may find the perceptual clustering of the speech sounds directly from the AM. By splitting the AM into its symmetric and skew-symmetric parts, as shown in Fig. 2.21, new insights may be clarified via this "perceptual eigen space." The small skew-symmetric components, such as /fa/ versus /θa/, as shown in the lower half of Fig. 2.21, seem to show talker biases. The much larger and robust symmetric part of the AM defines distances between the sounds, reflecting the underlying perceptual space, with larger confusions indicating smaller distances. Analyzing the AM in this way, as shown in Fig. 2.23, is much more informative than the many previous attempts where production based features were forced on the data, in an ad-hoc manner (Miller and Nicely, 1955; Wang and Bilger, 1973).

Two final subsections discuss AI validity (2.7) and criticisms (2.8). These sections are important because they review and summarize the variety of opinions on the validity of AI theory. With only a few exceptions, these criticisms have found to be invalid and unwarranted. Furthermore, we must keep in mind that there are at least three different versions of the AI. The most detailed (and the original version) by Fletcher and Galt has been the least studied (Rankovic, 1997). The one failing of AI may be its over-prediction of the score (i.e., $s < 1 - \prod_k e_k$) (i.e., redundancy) for the case of multiple pass bands (Kryter, 1960). However, Fletcher and Galt's 1950 model has not yet been tested for this case, so it may be premature to draw this conclusion.

Section 3 moves on to the topic of *Intelligibility*, and the effects of word and sentence context. Much of this work was done by Boothroyd, in a series of classic papers. More recently is the important work of Bronkhorst, who has significantly extended context modeling.

We then briefly discussed coarticulation and described the work of Furui, who has shown that, at least in some cases, the CV is carried by a very short 10-ms segment of dynamic speech change. This leads us to the conclusion that the events in the CV and VC can be fundamentally different. These ideas are similar to the early work at Haskins Laboratory based on painted speech, by Frank Cooper and his many colleagues.

Finally in Section 3.4, we discuss the interesting work of VanPetten, who has explored the relative timing of high-level syntax processing. Her work may be interpreted as showing that possible meanings are parsed simultaneously as phones are identified, on a time scale that is less than 200 ms, and typically well *before* the word is resolved.

4.1   ASR VERSUS HSR

Automatic speech recognition has long been a dream. While ASR does work today, and is available commercially, it is impossibly fragile. It is sensitive to noise, slight talker variations, and even slight changes in the microphone characteristics. While the exact reasons for the lack of robustness are not known, it is clear that present day ASR works very differently from human speech recognition. HSR accurately characterizes speech sounds into categories early in the process, well before context has been utilized. The SNR ratio at the input to the HSR feature extraction processing is determined by cochlear filtering. HSR seems to use features, such as correlations in timing, onsets and spectral balance, that are typically averaged and aliased away by ASR systems. The lack of emphasis on a high-quality spectral analysis with a quality filterbank is also a major failing with most modern ASR systems.

Typically ASR uses properties of the speech signal that are not used by human listeners. The SNR is typically determined by a low-quality filter bank, or no filter bank at all, in many ASR systems. This leads to significant performance reduction in noise. In ASR systems in use today, the issues of human-like feature extraction have largely been ignored. ASR "features" such as MFCC are heuristically derived. Sadly, a theoretical and experimental foundation, based on human performance metrics, is seriously lacking in present day ASR. Rather ASR is based on HMM models, inspired by low-entropy language models. And even these language models perform poorly when compared to human language processing (Miller and Isard, 1963).

Articulation index theory predicts that as long as the AI is constant, $s(AI)$ is constant, and in this case, the long-term spectrum is not relevant. One may conclude that the recognition measure $s(AI)$ is *not* determined by the long-term spectrum of the speech, rather it is determined by in-band and across-band modulations in the speech spectrum, at the output of the cochlea. These components are related to the short term loudness, as modeled by Fletcher and Galt's AI theory.

To better understand how HSR works we need much more *articulation matrix* data in the presences noise, filtering, and temporal modifications. We need more experimental databases of both MaxEnt CV and VC sounds, from many talkers. From an analysis of these AM databases, using the methods of information theory, we should be able to quantify the events, which are the key information bearing elements. Only by a direct comparison of the AI spectrogram [i.e., $AI(t, f, SNR)$] and the AM(SNR) [i.e., $P_{s,b}(SNR)$], near SNRs corresponding to the group bifurcation points (i.e., SNR_g), will we be able to identify the events that make up the underlying variables of HSR.

It is likely that speech is not a chain of Cs and Vs, rather the information is carried in the VC and CV transitions (Furui, 1986). We need new experimental data and a detailed analysis of this problem. It is likely that such an analysis holds the answers.

In my view we need to stop thinking in the frequency domain, and think more of events as a time-frequency sequence of events. Events are both time- and frequency-local. It is cochlear filtering that gives us high-fidelity extraction, removing noise from the signal by filtering. Understand the physics behind cochlear function is critical to robust ASR systems. However, it is the temporal correlation of modulations across frequency, over neighboring critical bands, that seem to be the key to the event, and it is the cochlea that defines these modulations.

Strong evidence for a timing basis of events was found by Furui (1986), in his temporal truncation experiments, where the recognition accuracy change by 40% over a 10 ms truncation interval change. This same time region was also found to be where the spectrum was changing most rapidly. It is difficult to reconcile these data with spectral measures, such as relative formant frequencies. The work of Robert Shannon and his colleagues also strongly supports a timing view (Shannon *et al.*, 1995) of feature support.

Randomizing words in meaningful sentences (reversing the word order, for example) only slightly reduces human performance. Furthermore these reductions may likely come from memory overload effects, not from reductions in processing strategy. Markov model based context processing, on the other hand, depends very strongly on word order, looking for a particular state-sequence of sounds or words. As predicted in 1963 by Miller and Isard, robustness in HSR is not a problem of language context effects.

If it is the errors in each event that are the source of the phone errors, as implied by the analysis of Miller and Nicely, and if the context processing can be treated as a noiseless state machine, as implied by the AI analysis of Fletcher, and later Boothroyd, then it follows that the problem of robustness in speech perception, to noise and filtering, will be solved if we can robustly extract the events from the speech by signal processing means. To do this we need to accurately identify and characterize events using the AM, after carefully controlling for context effects.

References

Abramson, A. S. and Lisker, L. (1970). Discrimiuability along the voicing continuum: Cross-language tests. In: *Proceedings of the Sixth International Congress of Phonetic Science*. Academia, Prague, pp. 569–573.

Allen, J. B. (1994). How do humans process and recognize speech? *IEEE Trans. Speech Audio*, **2**(4):567–577. doi:10.1109/89.326615

Allen, J. B. (1996). Harvey Fletcher's role in the creation of communication acoustics. *J. Acoust. Soc. Am.*, **99**(4):1825–1839.

Allen, J. B. (2001). Nonlinear cochlear signal processing. In: Jahn, A. and Santos-Sacchi, J., editors, *Physiology of the Ear, 2nd edition*. Singular Thomson Learning, San Diego, CA, Ch. 19, pp. 393–442.

Allen, J. B. (2005). Consonant recognition and the articulation index. *J. Acoust. Soc. Am.*, 117(4):2212–2223.

Allen, J. B. and Neely, S. T. (1997). Modeling the relation between the intensity JND and loudness for pure tones and wide-band noise. *J. Acoust. Soc. Am.*, **102**(6):3628–3646. doi:10.1121/1.420150

Baird, J. C. and Noma, E. (1978). *Fundamentals of Scaling and Psychophysics*. Wiley, New York. http://auditorymodels.org/jba/BOOKS_Historical/Baird/.

Bashford, J. A., Warren, R. M., and Lenz, P. W. (2000). Relative contributions of passband and filter skirts to the intelligibility of bandpass speech: Some effects of context and amplitude. *Acoust. Res. Lett. Online*, **1**(2):31–36. doi:10.1121/1.1329836

Beranek, L. L. (1947). The design of speech communication systems. *Proc. of the IRE*, **35**(9):880–890.

Boothroyd, A. (1968). Statistical theory of the speech discrimination score. *J. Acoust. Soc. Am.*, **43**(2):362–367. doi:10.1121/1.1910787

Boothroyd, A. (1978). Speech preception and sensorineural hearing loss. In: Studebaker, G. A. and Hochberg, I., editors, *Auditory Management of Hearing-Impaired Children: Principles and Prerequisites for Intervention*, University Park Press, Baltimore, pp. 117–144.

Boothroyd, A. (1993). Speech preception, sensorineural hearing loss, and hearing aids. In: Studebaker, G. A. and Hochberg, I., editors, *Acoustical Factors Affecting Hearing Aid Performance*, Allyn and Bacon, Boston, pp. 277–299.

Boothroyd, A. (2002). Context effects in spoken language perception. In *Proc. Congreso Internacional de Foniatra, Audiologa, Logopedia y Psicologa del lenguaje.*, pages 1–21, Universidad Pontificia de Salamanca. Salamanca, Spain. International Conf. on Foniatry, Audiology, Logopedics and Psycholinguistic.

Boothroyd, A. and Nittrouer, S. (1988). Mathematical treatment of context effects in phoneme and word recognition. *J. Acoust. Soc. Am.*, **84**(1):101–114.

Bronkhorst, A. W., Bosman, A. J., and Smoorenburg, G. F. (1993). A model for context effects in speech recognition. *J. Acoust. Soc. Am.*, **93**(1):499–509.

Bronkhorst, A. W., Brand, T., and Wagener, K. (2002). Evaluation of context effects in sentence regonition. *J. Acoust. Soc. Am.*, **111**(6):2874–2886. doi:10.1121/1.1458025

Campbell, G. (1922). Physical theory of the electric wave filter. *Bell System Tech. Jol.*, **1**(1):1–32.

Campbell, G. (1937). *Introduction to the Collected Papers of George Ashley Campbell.* AT&T. Archives, Warren, NJ.

Campbell, G. A. (1910). Telephonic intelligibility. *Phil. Mag.*, **19**(6):152–9.

Chisolm, T., Boothroyd, A., Reese, J., Roberts, R., Carlo, M., and Tarver, K. (2002). Invariance of linguistic context use across modality. *Assoc. for Resh. in Otolaryngology – Abstracts*, Abstract 197. http://www.aro.org/archives/2002/2002197.html.

Cover, T. M. and Thomas, J. A. (1991). *Elements of Information Theory.* John Wiley, New York.

Duggirala, V., Studebaker, G. A., Pavlovic, C. V., and Sherbecoe, R. L. (1988). Frequency importance functions for a feature recognition test material. *J. Acoust. Soc. Am.*, **83**(6):2372–2382.

Dunn, H. K. and White, S. D. (1940). Statistical measurements on conversational speech. *J. Acous. Soci. Am.*, **11**:278–288. doi:10.1121/1.1916034

Durlach, N. I., Braida, L. D., and Ito, Y. I. (1986). Towards a model for discrimination of broadband signals. *J. Acoust. Soc. Am.*, **80**(1):63–72.

Egan, J. P. and Wiener, F. M. (1946). On the intelligibility of bands of speech in noise. *J. Acoust. Soc. Am.*, **18**(2):435–441. doi:10.1121/1.1916384

Fechner, G. (1966). *Elemente der psychophysik.* Transl: in Adler, H., editor, *Elements of Psychophysics, Volume I.* Holt, Rinehart, and Winston, New York.

Flanagan, J. (1965). *Speech Analysis Synthesis and Perception.* Academic Press, New York.

Fletcher, H. (1921). An empirical theory of telephone quality. *AT&T Internal Memorandum*, **101**(6).

Fletcher, H. (1922a). The nature of speech and its interpretation. *J. Franklin Inst.*, **193**(6):729–747. doi:10.1016/S0016-0032(22)90319-9

Fletcher, H. (1922b). The nature of speech and its interpretation. *Bell System Tech. Jol.*, **i**:129–144.

Fletcher, H. (1929). *Speech and Hearing.* D. Van Nostrand Company, New York.

Fletcher, H. (1938). The mechanism of hearing as revealed through experiments on the masking effect of thermal noise. *Proc. Nat. Acad. Sci.*, **24**:265–274.

Fletcher, H. (1940). Auditory patterns. *Rev. Modern Phys.*, **12**(1):47–65. doi:10.1103/RevModPhys.12.47

Fletcher, H. (1953). *Speech and Hearing in Communication.* Robert E. Krieger Publ, Huntington, New York.

Fletcher, H. (1995). Speech and hearing in communication. In: Allen, J. B., editor, *The ASA Edition of Speech and Hearing in Communication.* Acoustical Society of America, New York.

Fletcher, H. and Galt, R. (1950). Perception of speech and its relation to telephony. *J. Acoust. Soc. Am.*, **22**:89–151. doi:10.1121/1.1906605

Fletcher, H. and Munson, W. (1937). Relation between loudness and masking. *J. Acoust. Soc. Am.*, **9**:1–10. doi:10.1121/1.1915904

French, N. R. and Steinberg, J. C. (1947). Factors governing the intelligibility of speech sounds. *J. Acoust. Soc. Am.*, **19**:90–119. doi:10.1121/1.1916407

Fry, T. (1928). *Probability and its Engineering Uses*. D. Van Nostrand, Princeton, NJ.

Furui, S. (1986). On the role of spectral transition for speech perception. *J. Acoust. Soc. Am.*, **80**(4):1016–1025.

Galt, R. H. (1940–1946). Speech AI studies: R. H. Galt Lab notebook 16. In: Rankovic, C. and Allen, J. B., editors, *Bell Labs Archive on Speech and Hearing, The Fletcher Years*. ASA, New York, p. 158.

Goldstein, L. (1980). Bias and asymmetry in speech perception. In: Fromkin, V. A., editor, *Errors in Linguistic Performance*. Academic Press, New York, Ch. 17, pp. 241–261.

Grant, K. and Seitz, P. (2000). The recognition of isolated words and words in sentences: Individual variability in the use of sentence context. *J. Acoust. Soc. Am.*, **107**(2):1000–1011. doi:10.1121/1.428280

Grant, K. W. and Seitz, P. F. (2001). The recognition of isolated words and words in sentences: Individual variability in the use of sentence context. *J. Acoust. Soc. Am.*, **107**(2):1000–1011. doi:10.1121/1.428280

Green, D. M. and Swets, J. A. (1966). *Signal Detection Theory and Psychophysics*. John Wiley, New York.

Green, D. M. and Swets, J. A. (1988). *Signal Detection Theory and Psychophysics*. Peninsula Publishing, Los Altos, CA. Reprint of Green and Swets, 1966.

Greenwood, D. D. (1990). A cochlear frequency-position function for several species—29 years later. *J. Acoust. Soc. Am.*, **87**(6):2592–2605.

Hartley, R. (1928). Transmission of information. *Bell System Tech. Jol.*, **3**(7):535–563.

Hartley, R. and Fry, T. C. (1921). The binaural location of pure tones. *Phy. Rev.*, **18**:431–442. doi:10.1103/PhysRev.18.431

Hartmann, W. M. (1997). *Signals, Sound, and Sensation*. AIP Press, Woodbury, NY.

Hirsh, I. J. A. (1954). Intelligibility of different speech materials. *J. Acoust. Soc. Am.*, **26**(4):530–538. doi:10.1121/1.1907370

Jakobson, R., Fant, C. G. M., and Halle, M. (1967). *Preliminaries to Speech Analysis: The Distinctive Features and Their correlates*. MIT Press, Cambridge, MA.

Kryter, K. D. (1960). Speech bandwidth compression throught spectrum selection. *J. Acoust. Soc. Am.*, **32**(5):547–556. doi:10.1121/1.1908140

Kryter, K. D. (1962a). Methods for the calculation and use of the articulation index. *J. Acoust. Soc. Am.*, **34**(11):1689–1697. doi:10.1121/1.1909094

Kryter, K. D. (1962b). Validation of the articulation index. *J. Acoust. Soc. Am.*, **34**(11):1698–1702. doi:10.1121/1.1909096

Lippmann, R. P. (1996). Accurate consonant perception without mid-frequency speech energy. *IEEE Trans. Speech Audio Process.*, **4**:66–69. doi:10.1109/TSA.1996.481454

Miller, G. A. (1947a). The masking of speech. *Psychol. Bull.*, **44**(2):105–129.

Miller, G. A. (1947b). Sensitivity to changes in the intensity of white noise and its relation to masking and loudness. *J. Acoust. Soc. Am.*, **19**:609–619. doi:10.1121/1.1916528

Miller, G. A. (1951). *Languange and Communication*. McGraw Hill, New York.

Miller, G. A. (1962). Decision units in the perception of speech. *IRE Transactions on Information Theory*, **81**(2):81–83, http://www.cisp.org/imp/march2001/03_01miller.htm. doi:10.1109/TIT.1962.1057697

Miller, G. A. (2001). Ambiguous words. *iMP Mag.*, pp. 1–10.

Miller, G. A., Heise, G. A., and Lichten, W. (1951). The intelligibility of speech as a function of the context of the test material. *J. Exp. Psychol.*, **41**:329–335.

Miller, G. A. and Isard, S. (1963). Some perceptual consequences of linguistic rules. *J. Verbal Learn. Verbal Behav.*, **2**:217–228. doi:10.1016/S0022-5371(63)80087-0

Miller, G. A. and Mitchell, S. (1947). Effects of distortion on the intelligibility of speech at high altitudes. *J. Acoust. Soc. Am.*, **19**(1):120–125. doi:10.1121/1.1916408

Miller, G. A. and Nicely, P. E. (1955). An analysis of perceptual confusions among some english consonants. *J. Acoust. Soc. Am.*, **27**(2):338–352. doi:10.1121/1.1907526

Miller, G. A., Wiener, F. M., and Stevens, S. S. (1946). Combat instrumentation. II. Transmission and reception of sounds under combat conditions. Summary Technical Report of NDRC Division 17.3. NDRC (goverment), Washington, DC.

Müsch, H. (2000). Review and computer implementation of Fletcher and Galt's method of calculationg the Articulation Index. *Acoust. Res. Lett. Online*, **2**(1):25–30. doi:10.1121/1.1346976

Musch, H. and Buus, S. (2001a). Using statistical decision theory to predict speech intelligibility I. Model structure. *J. Acoust. Soc. Am.*, **109**(6):2896–2909. doi:10.1121/1.1371971

Musch, H. and Buus, S. (2001b). Using statistical decision theory to predict speech intelligibility II. Measurement and prediction of consonant-descrimination performance. *J. Acoust. Soc. Am.*, **109**(6):2910–2920.

Papoulis, A. (1965). *Probability, Random Variables, and Stochastic Processes.* McGraw–Hill, New York.

Parker, F. (1977). Distinctive features and acoustic cues. *J. Acoust. Soc. Am.*, **62**(4):1051–1054.

Pickett, J. M. and Pollack, I. (1958). Prediction of speech intelligibility at high noise levels. *J. Acoust. Soc. Am.*, **30**(10):955–963. doi:10.1121/1.1909416

Rankovic, C. M. (1997). Prediction of speech reception for listeners with sensorineural hearing loss. In Jesteadt, W. *et al.*, editors, *Modeling Sensorineural Hearing Loss.* Lawrence Erlbaum, Mahwah, NJ, pp. 421–431.

Rankovic, C. M. (2002). Articulation index predictions for hearing-impaired listeners with and without cochlear dead regions (l). *J. Acoust. Soc. Am.*, **111**(6):2545–2548. doi:10.1121/1.1476922

Rankovic, C. and Allen, J. B. (2000). Study of speech and hearing at Bell Telephone Laboratories: The Fletcher years; CDROM containing Correspondence Files (1917–1933), Internal reports and several of the many Lab Notebooks of R. Galt. Acoustical Society of America, Melville, New York.

Rayleigh, L. (1908). Acoustical notes—viii. *Philosoph. Mag.*, **16**(6):235–246.

S3.5-1997 (1997). *Methods for Calculation of the Speech Intelligibility Index (SII-97)*. American National Standards Institute, New York.

Shannon, C. E. (1948). The mathematical theory of communication. *Bell Syst. Tech. Jol.*, **27**:379–423 (parts I, II), 623–656 (part III).

Shannon, C. E. (1951). Prediction and entropy of printed English. *Bell System Tech. Jol.*, **30**:50–64.

Shannon, R. V., Zeng, F. G., Kamath, V., Wygonski, J., and Ekelid, M. (1995). Speech recognition with primarily temporal cues. *Science*, **270**:303–304.

Shepard, R. (1972). Psychological representation of speech sounds. In David, E. and Denies, P., editors, *Human Communication: A Unified View*. McGraw–Hill, New York, Ch. 4, pp. 67–113.

Shepard, R. N. (1974). Representation of structure in similarity data: Problems and prospects. *Psychometrica*, **39**:373–421.

Shera, C. A., Guinan, J. J., and Oxenham, A. J. (2002). Revised estimates of human cochlear tuning from otoacoustic and behavioral measurements. *Proc. Natl. Acad. Sci. USA*, **99**:3318–3323. doi:10.1073/pnas.032675099

Siebert, W. (1970). Frequency discrimination in the auditory system: Place or periodicity mechanisms? *Proc. IEEE*, pp. 723–730.

Sinex, D., McDonald, L., and Mott, J. (1991). Neural correlates of nonmonotonic temporal acuity for voice onset time. *J. Acoust. Soc. Am.*, **90**(5):2441–2449.

Soli, S. D., Arabie, P., and Carrol, J.D. (1986). Discrete representation of perceptual structure underlying consonant confusions. *J. Acoust. Soc. Am.*, **79**(3):826–837.

Steeneken, H. and Houtgast, T. (1980). A physical method for measuring speech-transmission quality. *J. Acoust. Soc. Am.*, 67(1):318–326.

Steeneken, H. and Houtgast, T. (1999). Mutual dependence of octave-band weights in predicting speech intelligibility. *Speech Commun.*, **28**:109–123. doi:10.1016/S0167-6393(99)00007-2

Stevens, H. and Wickesberg, R. E. (1999). Ensemble responses of the auditory nerve to normal and whispered stop consonants. *Hearing Res.*, **131**:47–62. doi:10.1016/S0378-5955(99)00014-3

Stevens, H. and Wickesberg, R. E. (2002). Representation of whispered word-final stop consonants in the auditory nerve. *Hearing Res.*, **173**:119–133. doi:10.1016/S0378-5955(02)00608-1

Stevens, S. and Davis, H. (1938). *Hearing, Its Psychology and Physiology*. Republished by the the Acoustical Society of America in 1983, Woodbury, New York.

Van Petten, C., Coulson, S., Rubin, S., Planten, E., and Parks, M. (1999). Time course of word identification and semantic integration in spoken language. *J. of Exp. Psych.: Learning, Memory and Cognition*, 25(2):394–417. doi:10.1037//0278-7393.25.2.394

Van Petten, C. and Kutas, M. (1991). Electrophysiological evidence for the flexibility of lexical processing. In Simpson, G. B., editor, *Understanding Word and Sentence; Advances in Psychology 77*, chapter 6, pages 129–174. North-Holland, Amsterdam.

Van Valkenburg, M. (1964). *Network Analysis, 2nd* edition. Prentice-Hall, Englewood Cliffs, NJ.

Wang, M. D. and Bilger, R. C. (1973). Consonant confusions in noise: A study of perceptual features. *J. Acoust. Soc. Am.*, **54**:1248–1266.

Warren, R. M., Riener, K. R., Bashford, J. A., and Brubaker, B. S. (1995). Spectral redundancy: Intelligibility of sentences heard through narrow spectral slits. *Percep. Psychophys.*, **57**:175–182.

Wozencraft, J. M. and Jacobs, I. M. (1965). *Principles of Communication Engineering*. John Wiley, New York.

The Author

JONT B. ALLEN

Jont B. Allen received his B.S. in electrical engineering from the University of Illinois, Urbana-Champaign, in 1966, and his M.S. and Ph.D. in electrical engineering from the University of Pennsylvania in 1968 and 1970, respectively. After graduation he joined Bell Laboratories, and was in the Acoustics Research Department in Murray Hill, NJ, from 1974–1996, as a distinguished member of technical staff. Since 1996 Dr. Allen has been a Technology Leader at AT&T Labs-Research. Since Aug. 2003, Allen is an Associate Professor in ECE, at the University of Illinois, and on the research staff of the Beckman Inst., Urbana IL.

During his 32 year AT&T Bell Labs (and later AT&T Labs) career Dr. Allen has specialized in auditory signal processing. In the last 10 years he has concentrated on the problem of human speech recognition. His expertise spans the areas of signal processing, physical acoustics, cochlear modeling, auditory neurophysiology, auditory psychophysics, and human speech recognition.

Dr. Allen is a fellow (May 1981) of the Acoustical Society of America (ASA) and fellow (January 1985) of the Institute of Electrical and Electronic Engineers (IEEE). In 1986 he was awarded the IEEE Acoustics Speech and Signal Processing (ASSP) Society Meritorious Service Award, and in 2000 received an IEEE Third Millennium Medal.

He is a past member of the Executive Council of the ASA, the Administration Committee (ADCOM) of the IEEE ASSP, has served as Editor of the ASSP Transactions, as Chairman of the Publication Board of the ASSP Society, as General Chairman of the International Conference on ASSP (ICASSP-1988), and on numerous committees of both the ASA and the ASSP. He is presently a member of ASA Publications Policy Board. He has organized several workshops and conferences on hearing research and signal processing. In 1984 he received funding from NIH to host the 2^d International Mechanics of Hearing Workshop. He has a strong interest in electronic publications and has produced several CDROM publications, including suggesting and then overseeing technical details of the publication of the *J. Acoust. Soc. Am.* in DjVu format, and developed the first subject classification system for the IEEE Transactions of the ASSP, as well as the ASSP Magazine.

In 1986–88 Dr. Allen participated in the development of the AT&T multi-band compression hearing aid, later sold under the ReSound and Danavox name, and served as a member of the ReSound and SoundID Scientific advisory boards. In 1990 he was an Osher Fellow at the Exploratorium museum in San Francisco. In 1991–92 he served as an International Distinguished Lecturer for the IEEE Signal Processing Society. In 1993 he served on the Dean's Advisory Council at the University of Pennsylvania. In 1994 he spent 5 weeks as Visiting Scientist and Distinguished Lecturer at the University of Calgary. Since 1987 he has been an Adjunct Associate Research Scientist in the Department of Otolaryngology at Columbia University, and on the CUNY speech and Hearing Faculty (Adjunct). In 2000 he received the IEEE Millennium Award, and in 2004, an IBM faculty award. Dr. Allen has more than 90 publications (36 peer reviewed) and 16 patents in the areas of speech noise reduction, speech and audio coding, and hearing aids.